A PRACTICAL HANDBOOK: SPECTROMETER AND INTERFEROMETER DESIGN WITH ZEMAX OPTICSTUDIO

Muddasir Naeem
Tayyab Imran

Copyright © 2024 Muddasir Naeem

All rights reserved

The characters and events portrayed in this book are fictitious. Any similarity to real persons, living or dead, is coincidental and not intended by the author.

No part of this book may be reproduced, or stored in a retrieval system, or transmitted in any form or by any means, electronic, mechanical, photocopying, recording, or otherwise, without express written permission of the publisher.

ISBN-13: 9798339838692

Cover design by: Tayyab
Library of Congress Control Number: 2018675309
Printed in the United States of America

I want to dedicate this book to Allah, the originator of light "Prophet Muhammad (PBUH)," and my parents and sisters. To all the curious minds who strive to illuminate the unseen and unravel the mysteries of light, this work is dedicated to you. May it serve as a beacon in your quest for knowledge, guiding your exploration and inspiring you to push the boundaries of understanding. Let this book be a stepping-stone on your journey, encouraging you to discover the optical world's hidden wonders and share your insights with others who seek the same.

CONTENTS

Title Page

Copyright

Dedication

Foreword

Introduction

Preface

1. Spectrometry and Interferometry	1
References	17
2. Getting Started with ZEMAX	20
References	53
3. Design Simulation of Optical Spectrometer	55
Reference	75
4. Design simulation of Raman spectrometer	78
Reference	91
5. Design and Simulation of Interferometers	93
References	110
Appendix	114
Index	117
Acknowledgement	139
About The Author	141

About The Author	143
Books By This Author	145

FOREWORD

The journey of this book began with a shared passion for exploring the intricate world of optics and photonics. It was not just a collaboration between a mentor and a student but a confluence of curiosity, dedication, and the relentless pursuit of knowledge. Muddasir Naeem, then a bright and inquisitive student, embarked on this project with a keen interest in optical system design. Under my guidance, we delved into the spectrometer and interferometer design complexities, seeking to bridge the gap between theoretical concepts and practical applications.

As an educator and laser optics researcher, I have always believed that knowledge must be shared and expanded upon collaboratively. Muddasir's enthusiasm and determination to understand every nuance of ZEMAX OpticStudio—a powerful tool in the field of optics—matched this belief perfectly. Together, we navigated the challenges of translating complex optical phenomena into comprehensible models and simulations. It was through this synergy that the idea of creating a practical handbook emerged.

This book culminates countless hours of research, simulations, and discussions. It is designed to serve as a guide for students, researchers, and professionals who aspire to master the art of optical system design. Through step-by-step tutorials and real-world examples, we aim to empower readers to explore and create with confidence.

Muddasir's journey from student to author is a testament to the power of mentorship and collaboration. It has been a privilege to witness his growth and to contribute to a work that we hope will illuminate the paths of many others in the field of optics.

Tayyab Imran
Magurele, Romania

INTRODUCTION

The field of optics, encompassing the behavior and properties of light, has long played a pivotal role in scientific discovery, industrial applications, and technological innovation. In this context, **spectrometers** and **interferometers** are among the most important tools for analyzing light, with applications ranging from materials science to astronomy. However, designing and optimizing these complex optical systems is a challenging task requiring a deep understanding of optical principles and proficiency in advanced simulation tools.

This book, **"A Practical Handbook: Spectrometer and Interferometer Design with ZEMAX OpticStudio,"** is a comprehensive resource for anyone involved in optical system design. The book is a practical guide to creating and simulating spectrometers and interferometers using ZEMAX OpticStudio, a leading optical design software. It bridges the gap between theoretical optics and real-world applications by providing clear, step-by-step instructions for designing these systems.

The primary aim of this book is to offer a structured and practical approach to optical system design, focusing on spectrometers and interferometers. It is designed for students, researchers, and professionals in physics, engineering, and optics who seek to deepen their knowledge of optical system modeling and optimization.

The book emphasizes hands-on learning through **ZEMAX OpticStudio**, a powerful tool widely used in the optics and photonics industries. By guiding readers through the design and simulation of optical systems, the authors aim to make complex concepts more accessible and applicable to real-world scenarios. It covers sequential and non-sequential ray tracing, essential techniques for understanding how light interacts with different components in an optical system.

This book is invaluable for those working in or aspiring to work in fields where optical instruments are critical. Whether the reader is interested in improving imaging systems or designing innovative optical instruments, the book provides the tools and knowledge to tackle these challenges effectively. Moreover, **ZEMAX OpticStudio** is a widely adopted software in the optics industry, making this book a learning tool and a resource that enhances the professional skills needed in optical engineering, photonics, and scientific research.

In a world where optical technologies are becoming increasingly important in telecommunications, medical diagnostics, and space exploration, the ability to design, simulate, and optimize optical systems is more crucial than ever. This book is a vital resource for mastering these skills, offering readers a blend of theoretical insights and practical expertise. Through this book, the authors aim to inspire and equip the next generation of optical engineers and researchers to push the boundaries of what is possible in the fascinating world of light.

PREFACE

Optical system design is a complex and fascinating field that bridges the gap between theoretical physics and practical engineering. With technological advancements, tools like ZEMAX OpticStudio have revolutionized the way we approach optical design, simulation, and analysis. Our publications on ZEMAX-based design simulations of spectrometers and interferometers have played a key role in shaping this book, providing valuable insights into the design process. This book aims to offer a comprehensive guide for designing spectrometers and interferometers using ZEMAX. Whether you are a student, researcher, or professional, it will serve as a practical resource, featuring step-by-step tutorials and real-world examples to help you master optical system design.

The book is organized into five chapters, each focusing on a different aspect of optical system design using ZEMAX. Chapter 1 provides a foundational understanding of the dual nature of light and the principles behind spectrometry and interferometry. It introduces the fundamental concepts and applications of optical spectrometers and interferometers, setting the stage for the more detailed design and simulation work in the following chapters. In Chapter 2, readers are introduced to the ZEMAX OpticStudio software, including its main features, user interface, and different modes of operation. The chapter offers practical guidance on navigating the software and setting up the foundation for designing complex optical systems. Chapters 3 and 4 delve into the design and simulation of optical and Raman spectrometers using ZEMAX OpticStudio. It covers the entire process, from conceptual design to the interpretation of simulation results. Readers will learn how to model and optimize the spectrometers for various applications. The final Chapter 5 explores the design and simulation of Mach-Zehnder and Michelson interferometers. It discusses the theory behind these devices, their practical applications, and how to simulate them in ZEMAX OpticStudio.

By the end of this book, readers will have a solid understanding of designing, simulating, and analyzing optical systems, especially spectrometers and interferometers, using ZEMAX

OpticStudio. This resource is intended to empower you to tackle real-world challenges in optical engineering with confidence and creativity.

Muddasir Naeem
Lahore, Pakistan

1. SPECTROMETRY AND INTERFEROMETRY

The origin of optical technology started back in 1200 BCE. Greek philosophers Aristotle and Plato developed different theories of the nature of light. When the center of scholarship moved to the Arab world, Alhazen conducted extensive studies in optics around 1000 AD. He explained the law of reflection, angle of incident, and reflection. He also gave a detailed description of the human eye. Franciscan Roger Bacon (1215-1294) is considered the first scientist who gave the initial idea of using lenses and predicted the possibility of combining the lenses for the telescope. Kepler, in 1611, discovered total internal reflection; he evolved the treatment of first-order optics for a thin lens system. Wille Broad Snell found the law of refraction in 1621, one of the great moments in optics. Snell opened the door to modern applied optics. Newton concluded that white light is made up of a mixture of colors, even though his work included wave and emission theories. However, he dismissed wave theory because it couldn't explain the straight-line propagation of light. Christiaan Huygens (1629-1695) extended the wave theory and explained the laws of reflection using wave theory. The weight of Newton's opinion hung like a shroud over wave theory during the eighteenth century, but Euler (1707-1783) continued working on wave theory; he explained that undesirable color effects in a lens were absent in the eye because of media present dispersion and proposed the construction of achromatic lenses. Wave theory of light was reborn at the hands of Thomas Young (1773-1829). 1845, Michael Faraday established an interrelationship between electromagnetism and light[1]. Albert Einstein 1905 postulated that light always propagates in space with a definite velocity c, independent of the emitting body[2]. The first laser was constructed in 1960, and within ten years, its range was expanded from infrared to ultraviolet.

1.1 Dual Nature Of Light

Light is not only a wave but also a particle. Light appears to have two different sets of behavior under different circumstances. Light has a dual nature; sometimes, it behaves as a particle and a wave.

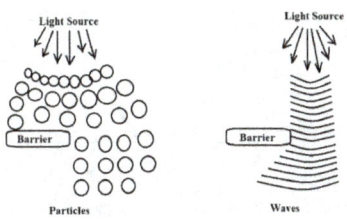

Figure 1.1 Dual nature of light

Early scientists believed light could only behave as a wave[1]. They explained that light is an electromagnetic wave through experiments on diffraction and interference[2]. However, Newton (1642-1727) rejected wave theory to emphasize emission theory. In 1801, Thomas Young performed a crucial experiment with interference[1]. He obtained dark and bright patterns called constructive and destructive interference, proving the **wave nature of light**, as shown in Figure 1.2.

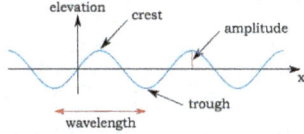

Figure 1.2 Wave nature of light

In 1905, Einstein proposed a theory that asserted light contains energy particles[1]. Einstein's photoelectric effect demonstrates that **light acts like a stream of particles**, known as photons, as illustrated in Figure 1.3.

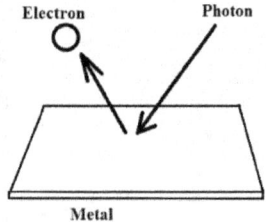

Figure 1.3 Particle nature of light

1.2 Spectrometry

With the invention of spectroscopy, there was a paradigm shift in optics[3]. Optical spectroscopy has extensively contributed to the field of physics over the last century. Spectroscopy has emerged as a powerful technique to uncover the wealth of galaxies, asteroids in the solar system, ion composition of distant stars, emission and absorption spectra of various materials, etc. The optical system's resolution is limited somehow, and the growing requirement to study at the nanoscale requires a fine-resolution spectrometer. Robert Wilhelm Bunsen and Robert Kirchoff, in 1859, invented the first spectroscope[1]. It was used to identify the materials when they emit light upon heating[4]. Several approaches have been proposed for spectrometers in the visible range, off-chip separation of wavelengths, grating spectrometers, and frequency-selective detection of light[5].

1.3 Optical Spectrometer

In the 1960s, lasers were available as sources for the excitation of samples, and a more sophisticated design of optical systems was started. A double-grating monochromator with a different fiber arrangement and an Offner system with a convex grating exceptional in aberration correction and distortion control[6] were proposed. Optical spectrometers, also called photometers, consist of one entrance slit through which light from the source enters. After a collimating mirror collimates that light, it falls on grating a dispersive element. Diffraction grating will disperse the light into its spectral components, which are focused by a focusing mirror and then detected by the CCD detector. The setup is connected to any automation software to form a graph and give spectral information to obtain the data. Optical spectrometers detect light intensity as a function of wavelength or frequency. Light is dispersed by using a prism or by using a diffraction grating. There are two types of optical spectrometers[7].

Optical emission spectrometers[8] are used to determine the composition of different elements in a sample accurately. First, a high-voltage spark is applied to the sample, ionizing the particle on the surface into a plasma. Then, the particles and ions emit photons of a particular wavelength, which are detected using the spectrometer. In an **Optical absorption spectrometer**, the light of the continuous spectrum from a source is applied to the sample. The sample absorbs only a particular wavelength, and the rest of the wavelength passes through it. The spectrometer detects the spectrum, and the absorbed wavelength will appear as a black line[9].

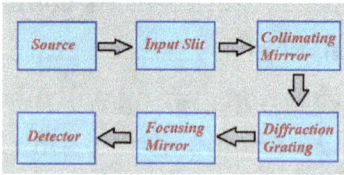

Figure 1.4 Block diagram of optical spectrometer

Major Components of Optical Spectrometer: Figure 1.4 shows the main parts in the block diagram of the optical spectrometer[10].

Entrance Slit: The entrance slit is also known as the input slit. This slit defines the portion of the light that will enter the optical bench and is essential for the spectrometer's performance. It is one of the factors leading to the determination of spectral resolution. The slit determines the throughput of the optical system by allowing some optical power into the system and blocking the rest percentage of optical power. The slit is an aperture that guides the beam into the spectrometer. The slit must be placed very accurately within the spectrometer's optical path; otherwise, it will compromise the calibration of the spectrometer. Aperture size has a relevant effect on optical uniqueness, such as spectral resolution. The most commonly used slit sizes in spectrometers are 10, 25, 50, 100, and 200 μm. Optical fiber cables are also employed as a method for light input, allowing light to travel through them and enter the spectrometer.

Collimating Mirror: The word collimate means parallel; the collimating mirror makes the beam parallel when the light from a source or input slit falls on it. Concave mirrors focus the light onto the detector, usually with a focal length of +300 mm. The collimating mirror collimates the light and takes the beam onto the grating.

Diffraction Grating: The diffraction grating disperses the light from the collimating mirror into different wavelengths at different diffraction angles. There are two types of gratings: transmission grating and reflection grating. The former type is not commonly used in optical spectroscopy because it has certain limitations in the absorption of dispersive material. Reflection gratings are further subdivided into plane and concave gratings. Optical resolution can be achieved with diffraction grating, and the wavelength range can be determined. The selection of correct grating is an essential factor in determining the best spectral results for applications. Dispersion depends on the number of grooves per mm ruled onto the grating. It is known as groove frequency or groove density. The wavelength coverage of the spectrometer is determined by the grating groove frequency, which is a critical factor in the spectral resolution. The wavelength coverage of a spectrometer has an inverse relation with the grating dispersion due to fixed geometry. For higher dispersion, the spectrometer will have a high-resolution power. Decreasing the groove frequency decreases the dispersion and increases wavelength coverage at the expense of spectral resolution. The groove facet angle, also known as the blaze angle, defines the shape of the diffraction curve. Blaze wavelength can also be calculated, determining which blaze angle corresponds to which peak.

Focusing mirror: The focusing mirror focuses the

light at one point. In the spectrometer, light dispersed from diffraction grating is focused by a focusing mirror. The dispersed light from the grating is focused onto the detector, which is placed at the focal point of the focusing mirror.

Detector: The detector gives the spectral information of the light coming to it. Usually, a charged coupled device (CCD) is used as a detector, but a complementary metal-oxide-semiconductor (CMOS) can also be used. A detector or radiation transducer is any device that converts an optical signal into a measurable phenomenon; then, these measurable phenomena will be converted into an impulse electrical signal. There should be a certain level of noise in detectors that can be acceptable; above that level, it will distort the signal. Noise is an unwanted signal from any source other than the signal of interest.

1.4 Raman Spectrometer

Raman Scattered Light: When light is incident on material, part of the light scatters, and part of a scattered portion contains the same wavelength as an incident, called Rayleigh scattering. Raman spectroscopy works on passing monochromatic light through a material to see if it is reflected, absorbed, or scattered. As the vibration and rotational properties alter, the scattered photons have a different frequency than the incident photon.

Raman spectroscopy is a method used to detect vibrational, rotational, and other states in the molecular system using scattered light. Raman spectroscopy work is based on the inelastic scattering of light. The Raman scattering phenomenon was first discovered in 1928 by Sir C.V. Raman while studying the scattering of sunlight filtered through transparent materials. He noticed that a small fraction of the scattered light had different frequencies than the incident light, indicating an energy exchange with the material's molecules. This observation led to the discovery of what is now known as the Raman effect[11]. Light interacts with molecular vibrations within the material, which change frequency when absorbed light is re-emitted by the material. Energy shift caused by these vibrational and sometimes rotational levels is measured in Raman spectroscopy.

If a molecule without Raman-active modes absorbs a photon with a frequency v_0, the excited molecule goes down to the same vibrational level.

It emits light as an excitation source of the same frequency υ_0. This type of interaction is called an elastic Rayleigh scattering.

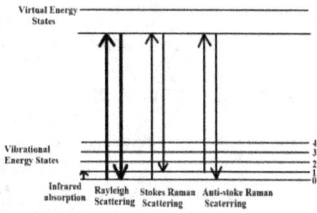

Figure 1.5 Schematic diagram of stokes and anti-stokes scattering in Raman spectroscopy

A photon of frequency υ_0 is absorbed through a Raman-active molecule in the simple vibrational state at the moment of contact. The photon's energy is transmitted to the Raman-active mode with frequency υ_m, and the corresponding frequency of the light dispersed is concentrated to $\upsilon_0 - \upsilon_m$. This Raman frequency is called Stokes frequency or simply "Stokes".

A photon with frequency υ_0 is in contact with a Raman-active molecule at the moment of interaction previously in the excited vibrational state. A large amount of energy is released from the excited Raman active mode, the molecule returns to its elementary vibrational state, and the resulting scattered light frequency goes up to $\upsilon_0 + \upsilon_m$. This type of Raman frequency is termed "Anti-Stokes".

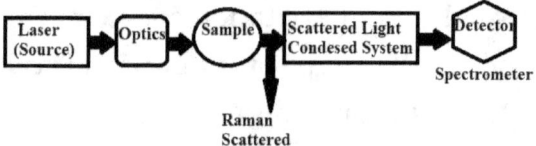

Figure 1.6 Basic configuration of Raman spectrometer

Essential components of Raman spectrophotometer

Source: Raman spectroscopy uses a laser light source with a small spectral width and high wavelength stability. A laser source is required because it is used to excite the target species. In a simple spectrometer, white light of different wavelengths is the source instead of a monochromatic source. This is the fundamental difference between the Raman spectrometer and the simple spectrometer.

Optics/Sample: With the help of a macro beam mirror and beam splitter, source light falls on the sample, and the filter gathers Raman scattered light(stokes) and filters out Rayleigh and anti-stoke light.

Spectrometer: When Raman scattered light enters the spectrometer and it falls on a diffraction grating, it bends the diffused light of Raman according to wavelength. The detector captures the signal and transfers it to the computer for decoding. Since scattered light from Raman is very faint, a detector with high sensitivity is needed.

1.5 Interferometry

Interferometry is an approach that extracts information from the interference of overlapping waves. Interferometry employs the idea of superposition to combine waves in such a manner that the outcome of their combination has some significant attribute, which is diagnostic of the original state of the waves. This works because when two waves with the same frequency combine, the resultant intensity pattern depends on the phase difference between the two waves— waves that are in phase will suffer constructive interference. In contrast, waves out of phase are prone to destructive interference. Waves that are neither totally in phase nor entirely out of phase will exhibit an intermediate intensity pattern, which may be used to calculate their relative phase difference and other information related to the source. Most interferometers employ light or another type of electromagnetic wave[12]. Different geometries of interferometers are shown in Figure 1.7.

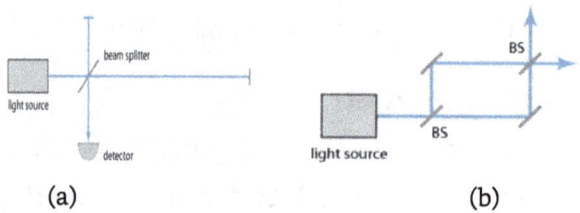

Figure 1.7 Schematic geometry of (a) Michelson interferometer (b) Mach-Zehnder interferometer

Interferometers are systems that use interference to extract information. They are frequently employed in research and industry to measure minute displacements, changes in refractive index, and surface imperfections. Most interferometers divide the light from a monochromatic light source into two beams that travel in distinct optical channels and are combined to generate interference; however, two incoherent sources can also interfere under certain conditions. The interference fringes provide details regarding the difference in optical path lengths. Interferometers are among the most accurate length-measuring tools in analytical research; they measure dimensions and the shape of optical elements with nanoscale precision. They are employed in Fourier transform spectroscopy to analyze light with absorption or emission patterns associated with a material or combination. An astronomy interferometer comprises two or more distinct telescopes that combine their signals to provide the resolution of a telescope with a diameter equal to the greatest separation between its component pieces. A beam splitter, a partially reflecting mirror, often splits a single incoming coherent light beam into two identical beams. Each of these rays follows a distinct path and undergoes recombination before arriving at a detector. The path difference, or distance traveled by each beam, causes a phase difference between them. The inserted phase difference causes the interference pattern generated by the initially identical waves. Suppose a single beam is divided into two pathways. In that case, the phase difference indicates anything that alters the phase along the routes, possibly due to a physical change in path length or a shift in refractive index over the path[13].

Interferometers are the highest-precision length-measuring tools used in analytical research; they measure lengths and the shape of optical components with nanoscale precision. They are used to detect gravitational waves and measure anything from the slightest alterations on the surface of a tiny creature to the structure of vast expanses of gas and dust in the distant Universe via radio interferometry. Laser interferometry has been used to estimate the diameter of circular cross-sectional fibers. The fiber is inserted into the beam, and the interference fringes are displayed on a screen[14].

REFERENCES

1. Hecht, E. (1998). Optics Addison Wesley Longman. Inc., New York, NY,.
2. Darrigol, O. (2012). A history of optics from Greek antiquity to the nineteenth century. OUP Oxford.
3. Gatkine, P., Zimerman, G., & Warner, E. (2018, September). A do-it-yourself spectrograph kit for educational outreach in optics and photonics. In Optics Education and Outreach V (Vol. 10741, pp. 185-191). SPIE.
4. Henderson, B. (1995). Optical spectrometers. Handbook of Optics, 2, 20-1.
5. Zhu, A. Y., Chen, W. T., Khorasaninejad, M., Oh, J., Zaidi, A., Mishra, I., ... & Capasso, F. (2017). Ultra-compact visible chiral spectrometer with metalenses. APL Photonics, 2(3).
6. Prieto-Blanco, X., Montero-Orille, C., González-Nuñez, H., Mouriz, M. D., Lago, E. L., & de la Fuente, R. (2010). The Offner imaging spectrometer in quadrature. Optics Express, 18(12), 12756-12769.
7. Naeem, M., Fatima, N. U. A., Hussain, M., Imran, T., & Bhatti, A. S.

(2022). Design Simulation of Czerny–Turner Configuration-Based Raman Spectrometer Using Physical Optics Propagation Algorithm. Optics, 3(1), 1-7.
8. Soltanpour, P. N., Jones Jr, J. B., & Workman, S. M. (1983). Optical emission spectrometry. Methods of Soil Analysis: Part 2 Chemical and Microbiological Properties, 9, 29-65.
9. Prieto-Blanco, X., Montero-Orille, C., Couce, B., & de la Fuente, R. (2008). Optical configurations for imaging spectrometers. Computational Intelligence for Remote Sensing, 1-25.
10. Naeem, M., Imran, T., Hussain, M., & Bhatti, A. S. (2022). Design simulation and data analysis of an optical spectrometer. Optics, 3(3), 304-312.
11. Raman, C. V. (1929). Part II.—The Raman effect. Investigation of molecular structure by light scattering. Transactions of the Faraday Society, 25, 781-792.
12. Brown, R. H. (1974). The intensity interferometer; its application to astronomy, Taylor & Francis, London.
13. Zetie, K. P., Adams, S. F., & Tocknell, R. M. (2000). How does a Mach-Zehnder interferometer work? Physics Education, 35(1), 46.

14. Mehra, R., & Tripathi, J. (2010). Mach zehnder Interferometer and it's Applications. International Journal of Computer Applications, 1(9), 110-118.

2. GETTING STARTED WITH ZEMAX

ZEMAX[1] is a software and services company founded in 1991. It offers design software for the optics industry. It is a powerful and widely used optical design software that plays an important role in optics and photonics. Zemax LLC developed a sophisticated platform for designing, analyzing, and optimizing optical systems. The versatility of ZEMAX makes it an essential tool for applications ranging from simple lens designs to complex imaging systems, spectrometers, and interferometers.

The main strength of ZEMAX is its intuitive graphical user interface, which allows users to create optical systems by designing and programming optical components, such as lenses, mirrors, and detectors, on a virtual workspace. The software uses a powerful ray tracing engine that simulates the actions of light as it interacts with these objects. It allows users to predict the light flow through the system, calculate performance, and optimize settings to achieve desired results. The ability of ZEMAX to handle both sequential and non-sequential beam tracing allows it to adapt to different optical systems.

What are sequential and non-sequential beam tracing modes? In **sequential mode**, light rays follow a predefined, ordered path through optical elements like lenses, mirrors, and apertures. Each ray is traced step-by-step through this sequence. This is ideal for designing and analyzing traditional optical systems, such as lenses, telescopes, and microscopes, where the light path is well-defined. In **non-sequential** mode, light rays are not restricted to a specific order. Rays can scatter, reflect, refract, or even split when they encounter optical components, potentially interacting with multiple surfaces in any order. It is used for complex systems like lighting, stray light analysis, or systems with multiple light sources and components where light can take various paths.

Additionally, ZEMAX offers a complete library of optics and packages, facilitating the accurate simulation of real-world situations. Users can analyze optical systems for parameters such as image quality, spot diagrams, and wavefront aberrations, ensuring designs meet stringent performance standards. The software also facilitates tolerance analysis, empowering designers to account for manufacturing variables and environmental conditions. With scripting capabilities, ZEMAX can be automated for batch processing, making it suitable for high-throughput optical design projects.

In addition to its programming and analysis features, ZEMAX provides a valuable optimization tool, allowing users to fine-tune the optical system parameters.

2.1 Main Features Of Zemax

Ray Tracing: One of the core features of this software is its ray tracing engine, which allows users to simulate the path of light rays through optical components. It enables predicting how light will behave within an optical system and optimizes designs for various optical parameters, including image quality, spot diagrams, and aberrations.

Geometric and Physical Optics: It supports geometric and physical optics simulations, providing a holistic approach to optical system analysis. Geometric optics[2,3] are suitable for quick, approximate designs, while physical optics simulations account for wavefront effects, diffraction[4], and interference, making them suitable for high-precision optical systems.

Tolerance Analysis: Tolerance analysis tools allow users to assess component tolerances' impact on system performance. It ensures that designs can meet performance specifications even in less-than-ideal conditions.

Optimization: ZEMAX provides optimization tools to fine-tune optical system parameters automatically. Optimization algorithms are used to achieve specific performance goals, such as minimizing aberrations or maximizing throughput. This feature significantly speeds up the design process and improves system performance.

Library of Optical Components: An extensive library of pre-defined optical components, materials, and surfaces is included. A wide range of lenses, mirrors, prisms, filters, and detectors make it easy to model real-world optical systems accurately.

Interoperability: Supports data exchange with other engineering and design software, including CAD programs. This interoperability ensures that optical designs can be easily incorporated into larger systems.

2.2 Opticstudio

It is used to design and analyze imaging and illuminance systems. It can model all optical elements and the effect of optical coating on the surface of elements. It has two modes, i.e., sequential and non-sequential modes, which we have already discussed. Sequential mode[5,6] is also suitable when analyzing the performance of imaging and afocal systems; Non-sequential mode is optimal for analyzing stray light, scattering, and illumination. In sequential mode, the propagated ray incident the optical element sequentially, i.e., the order in which optical elements are placed. In non-sequential mode, propagated rays can incident the optical elements in any order regardless of the order in which optical elements are placed. Some of the features available in the Zemax are explained below.

File Menu:

New: Restores Zemax back to the start-up condition.

Open: Open the existing Zemax file.

Save: Save the file.

Save as: Save the file with another name.

Use Session file: Toggles the environment setting between using and not using session files.

Back To Archives File: Saves the *.ZMX lens file and all associated files to a single *.ZAR archive file.

Restore From Archives File: Opens an existing Zemax archive file and restores all the individual files.

There are two distinct modes to the Zemax User Interface, as shown in screenshot

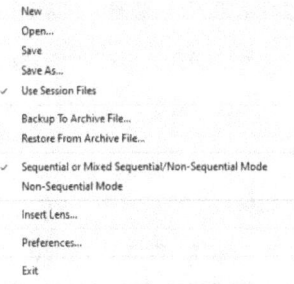

Sequential or Mixed Sequential/Non-Sequential Mode: All program capabilities are available in Sequential mode, including the option to place non-sequential groups on any surface. Zemax's Sequential mode traces rays that depart the object surface, including specified surfaces and non-sequential groups. This is the recommended method for building imaging systems or any system that requires optimization, tolerancing, and extensive picture analysis.

Non-Sequential Design (NSC mode): Non-Sequential mode[7] simplifies system analysis by repeatedly modifying the user interface and lens data.

Menus and button bars in Non-Sequential mode are simplified by removing unnecessary features. The fundamental analysis feature in the Non-Sequential mode is tracing rays to one or more detector objects. Using the Non-Sequential mode simplifies program operation on non-imaging devices. Moving from non-sequential to sequential mode is always possible without losing data. Switching from Sequential to Non-Sequential mode deletes all sequential data.

Insert Lens: Inserts surface data from a previously stored lens data to the current lens file

Preferences: Zemax allows users to set and save a variety of choices, which are then picked up when Zemax is started. The main configuration file is "Zemax.CFG"; deleting this file resets the start-up settings to their default state.

Exit: Exits Zemax. If the lens has been modified, Zemax will prompt you to save the lens. If not, the program exits.

Editor Menu:

Lens Data: The Lens Data Editor is the primary spreadsheet where most of the lens information will be input. The data comprises the radius of curvature, thickness, and glass for each surface in the system. Single lenses have two surfaces (front and back), and the object and image require one surface.

Merit Function: The Merit Function Editor defines, modifies, and reviews the system merit function.

Multi-Configuration: The Multi-Configuration Editor is quite similar to the Lens Data Editor. To modify a cell, move the pointer to it and enter new data. To set a solution on a cell, double-click with the left mouse button or use the solve type menu option.

Tolerance Data: The Tolerance Data editor sets, changes, and evaluates system tolerance settings.

Extra Data: The additional data editor functions similarly to the lens data spreadsheet editor but only displays and edits the extra data values.

System Menu:

Update: This option only changes the data in the Lens Data and Extra Data editors. Update recomputes first-order properties, pupil locations, semi-diameters, index data, and solves. This only affects data provided in the Lens Data and Extra Data editors.

Update All: Zemax's frequent recalculation of

MTF, ray fan, spot diagram, and other statistics when entering new data in the lens data editor would result in poor program reaction time. After making the necessary modifications to the lens, choose Update All to refresh and recompute all data windows. You may update individual visual and text windows (not editors) by double-clicking anywhere inside them.

General...: This option opens the General System Data dialog box, allowing you to provide common data for the lens as a whole rather than just a specific surface.

Fields...: The field dialog box allows you to specify field points. The buttons can turn field locations on or off and sort, store, and load data.

Wavelengths...: The wavelengths dialog box is used to specify wavelengths, weights, and the principal wavelength. The buttons allow wavelength activation, deactivation, and data sorting, saving, and loading. A list of regularly used wavelengths is also provided.

Next Configuration: This menu item quickly updates all visuals to match the next configuration (or zoom position). The whole spreadsheet, text, and graphics data will be updated.

Last Configuration: This menu item quickly updates all visuals to reflect the previous configuration (or zoom position). All spreadsheet, text, and graphical information will be updated.

2.3 Sequential Or Mixed Sequential/Non-Sequential Mode

Lens Data Editor (LDE) Window: A main component of the Zemax OpticStudio software for designing and analyzing optical systems. The Lens Data Editor window provides a graphical interface where users can input the parameters of optical components within an optical system. The LDE is a fundamental tool for configuring and fine-tuning optical designs within Zemax OpticStudio. Fig 2.1 shows the columns of different parameters necessary for designing an optical system, such as radius, thickness, semi-diameter, and conic. In LDE, the Surf: Type column is used to define the elements like source, detector, grating hologram, etc. The Radius column defines the radius of curvature of the element. If the element has a plane surface, the Radius will be infinite.

Thickness is the distance between the current element and the next one, or in the case of a thick element, it is the distance between the front and back surfaces. It does not define the thickness of the element, but glass defines the material of the element, such as BK7 and N-BK7. For elements to use as a mirror, the Glass type will be "mirror". Semi-diameter defines the size of the element.

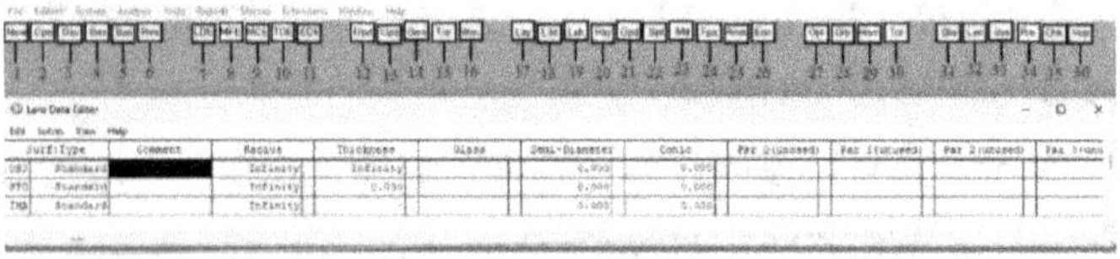

Figure 2.1 Lens Data Editor Window (LDE) in ZEMAX

1. New File	2. Open File
3. Save File	4. Save As
5. Backup to Archive File	6. Restore From Archive File
7. Lens Data Editor	8. Merit Function Editor
9. Multi-Config Editor	10. Tolerance Data Editor
11. Extra Data Editor	12. Update
13. Update All	14. General Lens Data
15. Field Data	16. Wavelength Data
17. 2D Layout	18. 3D Layout
19. Shaded Model Layout	20. Ray Fan
21. Opd Fan	22. Spot Diagram
23. Modulation Transfer Function	24. FFT PSF
25. RMS vs. Field	26. Diffraction Encircled Energy
27. Local Optimization	28. Global Optimization
29. Hammer Optimization	30. Tolerance
31. Glass Catalog	32. Lens Search
33. System Data	34. Prescription Data
35. System Check	36. Visual Optimization

Analysis Menu: The Analysis Menu contains thorough descriptions of each analysis feature that Zemax supports. In this sense, analysis refers to any graphical or textual data derived from data that defines the lens. This contains aberrations,

MTF, spot diagrams, and other calculations. Some of the analysis features are given below.

Layout:

2D Layout: This gives a YZ cross-section of the lens. This functionality is not accessible if you employ coordinate breaks, spider obscurations, obscuration decenters, X-angles, holograms, or other qualities that disrupt the rotational symmetry of the lens. Use the 3D layout instead.

3D Layout: Draws 3D layout plots of the lens system. Pressing the left, right, up, down, Page Up, or Page Down keys will rotate the displayed image for a different perspective.

Shaded Model: Draws a shaded solid representation of the lens using OpenGL graphics.

Zemax Element Drawing: This function generates a mechanical drawing of surface, singlet, cemented doublet, or cemented triplet components appropriate for optical shop manufacturing.

ISO Element Drawing: This function generates an ISO 10110 drawing of surface, singlet, or doublet components appropriate for optical shop manufacturing.

Fans:

Ray Aberration: Shows ray aberrations as a function of pupil coordinate.

Optical Path: Shows optical path difference as a function of pupil coordinate.

Pupil Aberration: Shows entrance pupil distortion as a function of pupil coordinate.

Spot Diagrams:

Standard: Show the spot diagrams. The maximal ray density depends on the number of shown fields, specified wavelengths, and available memory. Through-focus spot diagrams trace half the maximum number of rays as regular spot diagrams.

Through Focus: Show spot diagrams as they change with focal plane movements. Focus spot diagrams can help estimate astigmatism, analyze optimal focus, and determine the depth of focus.

Full Field: The spot diagram depicts all field points on a similar scale. The "Full Field" spot diagram differs from the "Standard" kind by plotting all spots using the same reference point rather than a distinct one for each field location.

Matrix: Display the spot diagram as a matrix of

separate diagrams, with each field along a row and wavelength along a column.

Configuration Matrix: Display the spot diagram as a matrix of separate diagrams, with each field on a row and each configuration down a column.

MTF:

FFT MTF: Computes the diffraction modulation transfer function (MTF)[8] data for all field points and is calculated using an FFT technique.

FFT Through Focus MTF: Computes FFT modulation transfer function data as a function of focus shift at a given frequency.

FFT Surface MTF: The FFT-computed MTF data is shown as a 3D surface, contour, greyscale, or false color map. This graphic helps see the MTF response for object orientations other than sagittal or tangential.

FFT MTF vs. Field: Computes FFT MTF data as a function of field location and plots it in a graph.

FFT MTF Map: The FFT MTF is calculated as a function of field location and shown across a rectangular portion of the field.

Huygens MTF: The diffraction modulation transfer function (MTF) data is calculated using the Huygens direct integration technique.

Huygens Through Focus MTF: The Huygens direct

integration approach is used to compute the diffraction modulation transfer function (MTF) data, which is then shown as a delta focus function.

Huygens Surface MTF: The Huygens direct integration technique is used to compute the diffraction modulation transfer function (MTF) data, which is then shown as a greyscale or false color graphic.

Huygens MTF vs. Field: Calculates the Huygens MTF data as a function of field location and shows it in a graph.

Geometric MTF[9]*:* Calculates the geometric MTF, an approximation of the diffraction MTF using ray aberration data.

Geometric Through Focus MTF: Computes the geometric MTF data by focusing on a certain spatial frequency.

Geometric MTF vs. Field: Determines the geometric modulation transfer function data as a function of field location. This feature is similar to the (diffraction) MTF vs. Field feature, except the geometric MTF is utilized instead of the diffraction-based MTF.

Geometric MTF Map: Calculates and shows geometric modulation transfer function data based on field location across a rectangular area.

PSF:

FFT PSF: The diffraction point spread function (PSF)[10] is calculated using the Fast Fourier Transform (FFT) technique.

FFT PSF Cross Section: This feature visualizes cross sections through the diffraction PSF.

FFT Line/Edge Spread: This feature plots edge or line spread functions by integrating the diffraction FFT PSF.

Huygens PSF: The diffraction PSF is calculated by directly integrating the Huygens wavelets approach. The Strehl ratio is also calculated.

Huygens PSF Cross Section: The diffraction PSF is calculated by directly integrating the Huygens wavelets approach. The Strehl ratio is also calculated.

Wavefront:

Wavefront Map: Displays the wavefront aberration.

Interferogram: Generates and displays interferograms.

Foucault Analysis: Generates and displays Foucault's knife-edge shadowgrams.

Surface:

Surface Sag: The sag of a surface is shown as a 2D color or contour map or as a 3D surface plot. This feature considers the size and geometry of any apertures on the surface, even those decentered. The sag is computed on a uniform grid of points in the XY plane, and its Z value is presented.

Surface Phase: Displays the phase of a surface as a 2D color or contour map or as a 3D surface plot.

Surface Curvature: Displays the tangential, sagittal, x, or y curvature of a surface as a 2D color or contour map or as a 3D surface plot.

Surface Sag Cross Section: Displays the sag of a surface as a cross-section.

Surface Phase Cross Section: Displays the phase of a surface as a cross-section.

Surface Curvature Cross Section: Shows the curvature of a surface in a cross-section.

RMS:

RMS vs. Field: Plots RMS radial, x, and y spot radius, RMS wavefront error, or Strehl ratio[11] vs field angle.

RMS vs. Wavelength: Plots RMS radial, x, and y spot radius, RMS wavefront error, or Strehl ratio vs wavelength.

RMS vs. Focus: Plots RMS radial, x, and y spot radius, RMS wavefront error, or Strehl ratio vs focus change.

RMS Field Map: Displays RMS radial, x, and y spot

radius, RMS wavefront error, or Strehl ratio as a function of field location in a rectangular map.

Encircled Energy:

Diffraction: The percentage of total energy contained varies based on distance from the primary ray or image centroid at a point object.

Geometric: Computes encircled energy using ray-image surface intercepts.

Geometric Line/Edge Spread: Calculates the geometric response of a line and an edge object.

Extended Source: Computes encircled energy using an extended source comparable to the geometric picture analysis feature.

Image Simulation: This feature mimics picture generation by combining a source bitmap file with an array of Point Spread functions. Diffraction, aberrations, distortion, relative lighting, picture orientation, and polarization are among the impacts under consideration.

Geometric Image Analysis: The geometric image analysis function has several uses. It may be used to model extended sources, examine resolution, describe object appearance, and offer insight into picture rotation. Image analysis is also beneficial for calculating multi-mode fiber coupling efficiency.

Geometric Bitmap Image Analysis: This feature generates an RGB color image based on ray tracing data, utilizing an RGB bitmap file as the source.

Light Source Analysis: This feature generates an image using ray tracing and a list of rays as the source.

Partially Coherent Image Analysis: This feature computes picture appearance by considering optical system diffraction, aberrations, and partial illumination coherence.

Extended Diffraction Image Analysis: This feature computes complicated diffraction picture attributes from extended sources, allowing for variations in the optical transfer function (OTF) over the field of view.

MA/BIM File Viewer: This feature displays IMA/BIM files.

Bitmap File Viewer: This function shows JPG, BMP, and PNG image files without processing.

Biocular Analysis:

Field of View: Displays the field of view for up to four configurations.

Divergence/Convergence: Divergence and convergence are displayed for binocular analysis.

Miscellaneous:

Field Curvature/Distortion: Displays the field

curvature and distortion plots.

Grid Distortion: Displays a grid of chief ray intercept points to indicate distortion.

Relative Illumination: Calculate the relative illumination as a function of the radial field coordinates for a uniform Lambertian scene. This feature also calculates the effective F/#.

Vignetting Plot: Calculates fractional vignetting as a function of field angle.

Footprint Diagram: Displays the beam's footprint overlaid on any surface. It is used to demonstrate vignetting and test surface apertures.

Longitudinal Aberration: The longitudinal aberration is shown as a function of pupil height at each wavelength.

Lateral Color: Displays the lateral color as a function of field height.

Y-Ybar Diagram: The Y-Ybar graphic plots the marginal and primary ray height for a paraxial skew ray at each lens surface.

Chromatic Focal Shift: Chromatic focal shift plot. This is a plot depicting the shift in back focal length in relation to the primary wavelength.

System Summary Graphic: Similar to the text-based system data report, this program displays a summary of the system data in a visual window.

Power Field Map: This feature calculates optical power or effective focal length as a function of field location. This characteristic is commonly used to analyze the spherical and cylinder power of Progressive Addition Lenses (PALs).

Power Pupil Map: This feature calculates optical power or effective focal length as a function of pupil location.

Incident Angle vs. Image Height: This feature calculates incidence angles for lower, chief and upper marginal rays based on picture height.

Aberration coefficients:

Seidel Coefficients: Displays Seidel (unconverted, transverse, and longitudinal), as well as wavefront aberration coefficients.

Seidel Diagram: Displays Seidel unconverted aberration coefficients as a bar chart.

Zernike Fringe Coefficients: Zernike coefficients are calculated using Fringe polynomials.

Zernike Standard Coefficients: This feature is identical to the "Zernike Fringe Coefficients" feature but uses a different numbering scheme. It allows for more fitting terms in the expansion, is orthogonal and normalized, and does not skip any terms.

Zernike Annular Coefficients: This feature is

identical to the "Zernike Standard Coefficients" feature; however, the pupil might be annular instead of round.

Zernike Coefficients vs. Field: Calculates and displays the Fringe, Standard, and Annular Zernike coefficients across the image's surface.

Calculations:

Ray Trace: Paraxial and real trace of a single ray.

Fiber Coupling Efficiency: This feature calculates the coupling efficiency for single-mode fiber coupling systems.

YNI Contributions: This feature displays each surface's paraxial YNI value, which is proportional to its Narcissus contribution.

Sag Table: This feature displays the sag (z-coordinate) of the selected surface at different Y-coordinate distances from the vertex. Compute the best fit spherical (BFS) radius and tabulate its sag and difference from the surface. The depth of material extracted from the BFS to create the asphere is also noted. The dimensionless slope (dz/dr) of the real surface and BFS are provided, along with their difference. This characteristic only considers the surface's Y-coordinate and should not be utilized if the surface is not rotationally symmetric.

Cardinal Points: This feature displays the positions of the focal, principal, anti-principal, nodal, and anti-nodal planes for a given surface and wavelength range. The computation is done for any defined wavelength and orientation (X-Z or Y-Z).

Glass and Gradient Index:

Dispersion Diagram: Plots the index of refraction vs wavelength for up to four distinct glasses.

Glass Map: Plots the names of glasses on a map based on their d-light index of refraction and Abbe V-numbers.

Athermal Glass Map: Plots the names of glasses on a map based on their chromatic and thermal power.

Internal Transmittance vs. Wavelength: Plots the internal transmittance of any thickness vs wavelength for up to four distinct glasses.

Dispersion vs. Wavelength: Plots dispersion versus wavelength for up to four distinct glasses.

Grin Profile: Plots the index of refraction along any axis for gradient index surfaces.

Gradium™ Profile: Plots the axial index profile of a Gradium surface.

Polarization:

Polarization Ray Trace: The polarization ray tracing feature presents a text window with all the polarization data for a single beam.

Polarization Pupil Map: Creates a graph of the polarization ellipse as a function of pupil location. This helps see the shift in polarization over the pupil.

Transmission: Calculates integrated and surface-by-surface transmission across an optical system with polarization effects.

Phase Aberration: Computes the polarization-induced phase aberration of an optical system.

Transmission Fan: Creates a graph of transmitted intensity by field and wavelength based on tangential or sagittal pupil fans.

Coatings: Computes the S, P, and average polarization intensity coefficients for reflection, transmission, absorption, Diattenuation, phase, and retardance for the specified surface as a function of incident angle and wavelength.

Physical Optics:

Paraxial Gaussian Beam: Calculates the parameters of a paraxial Gaussian beam. Paraxial Gaussian Beams are used for axial analysis of rotationally symmetric systems.

Skew Gaussian Beam: Calculates the Skew Gaussian Beam parameters. Skew beams can enter an optical system from any surface and move off-axis. Skew

Gaussian Beam parameters are estimated using actual rays, accounting for optical astigmatism but not higher-order aberrations.

Physical Optics Propagation: This feature propagates any coherent optical beam across the optical system. The analysis calculates the beam irradiance or phase on a plane tangent to the principal ray at the surface intersection. The propagation report can help establish the plane's orientation matrix. The irradiance or phase data appears after the beam refracts into or reflects off the end surface.

Beam File Viewer: This feature allows viewing and analysis of previously stored Zemax Beam Files (ZBF).

Source Viewers:

Directivity Plot: Displays a directivity diagram for IES, LDT, and RSMX source files.

Polar Plot: Displays a hemi- or full-sphere map of the source luminance for IES, LDT, and RSMX source files.

Source Illumination Map: Displays a realistic color picture of an illumination pattern created by one or more sources. The sources can be specified using IES, LDT, or RSMX source files.

Spectrum Plot: Displays the relative intensity vs.

wavelength for a defined spectral distribution.

Radiant Source Model™ Viewer: Displays source images contained within Radiant Source Model™ files.

CIE 1931/1976 Color Chart: These features display color charts based upon the CIE 1931 and 1976 color spaces.

Scatter Viewers:

Polar Plot: Displays BSDF as a function of radial and azimuthal angle on a 2D polar plot.

Scatter Function Viewer: Displays BSDF as a function of the magnitude of the scatter vector x.

Obsolete:

Wireframe: Draws a representation of the lens.

Solid Model: Draws a hidden-line representation of the lens.

Illumination XY Scan: Calculates relative illumination for the picture of an extended source projected along a line over any surface. This functionality is only suitable for systems that have a sufficiently decent picture of the object source.

Illumination 2D Surface: Calculates the relative illumination for an extended source on a 2D surface. This functionality is only suitable for systems that have a sufficiently decent picture of the object source.

Export IGES Line Work: Exports the current lens data as an IGES line work format file, with a few options.

Export 2D DXF File: Exports the current lens data to a 2D DXF file. It is only suitable for spherical symmetric lenses.

Export 3D DXF File: Exports the current lens data as a 3D DXF format file.

Conjugate Surface Analysis: This feature calculates the RMS spot radius on a surface using a point from a

preceding surface as the object. The feature deletes surfaces before and after conjugate surfaces and evaluates picture quality for a specified location on the new object surface.

2.4 Non-Sequential Mode

Imaging systems often use consecutive optical surfaces, with rays tracing from one surface to the next in a tight order. Each ray "hits" each surface just once in a preset order. The sequential model is efficient, practical, and comprehensive for several scenarios.

However, there are occasions when a non-sequential trace is necessary. Non-sequential implies that rays track in the physical order they impact objects or surfaces rather than the order indicated in the program interface. In non-sequential tracing, rays may repeatedly hit the same item while missing others. Rays often impact objects in a certain sequence based on their shape and the angle and location of the input ray. Non-sequential ray tracing benefits objects such as prisms, light pipes, lens arrays, reflectors, and Fresnel lenses. Certain analyses, such as stray or dispersed light effects, need a non-sequential setting. Previously, lens design algorithms that supported surfaces (rather than 3D objects) for sequential ray tracing used the same surface model for non-sequential ray tracing, where rays intersected surfaces in an out of sequence fashion. Using surfaces in a non-sequential method may not effectively represent many optical components. Lenses feature several surfaces, including front and rear, edges, and flattened exterior sides for mounting purposes. Surfaces not included in sequential surface ray tracing algorithms can intercept, refract, or reflect light. In complex prisms like a dove or roof prism, rays can cross many faces in a complicated order based on their input angle and location. The Non-Sequential mode removes unnecessary features from menus and button bars to simplify the user interface. Sequential ray tracing offers distinct features such as ray fans, MTF plots, and spot diagrams. Non-sequential mode analysis involves tracing rays to one or more detector objects. Using the Non-Sequential mode makes the software easier to use on non-imaging platforms. Some of the analysis features in non-sequential mode are described below.

LightningTrace Control: The LightningTrace Control launches and traces rays from a discrete mesh for analysis using defined sources, objects, and detectors. This function utilizes powerful ray interpolation technology to swiftly predict lighting patterns without tracing millions of rays. The program aims to determine approximate lighting patterns quickly, enabling faster setup and design of illumination systems than traditional ray tracing. When utilizing LightningTrace, a specific group of rays is fired and tracked from each specified source. The launch directions are determined along a mesh that surrounds the source. The Ray Sampling input determines the mesh resolution, which ranges from "Low (1X)" to "1024X". As the mesh size doubles, the number of mesh rays quadruples. Rays from each mesh start at the same location, corresponding to the source's center. LightningTrace uses a far-field description of each source rather than considering its size or spatial dispersion. LightningTrace analysis may be performed on Detector Rectangle, Color, and Polar objects. To observe results on these detectors, either the detector is a terminal detector (rays do not interact with other objects) or the material is set to "ABSORB". When examining results on a Detector Rectangle or Detector Color object, only spatial data is accessible. To retrieve angular data, use a Detector Polar object instead. The key advantage of LightningTrace is speed. Tracing rays from a mesh reduces the amount of rays utilized compared to traditional ray tracing methods. The system uses powerful ray interpolation technology to determine the behavior of mesh rays in between areas. LightningTrace provides an estimate of system performance, which may result in artifacts like hot spots.

Ray Database Viewer: The Ray Database Viewer reads a previously saved ZRD file and presents the results in text format.

Detector Viewer: The Detector Viewer shows data from detectors in either visual or text format.

After recording ray data on detectors or in a database, the Show Data option displays specific aspects of the data, as seen below:

Incoherent Irradiance: This is the amount of incoherent power per unit area that varies with the spatial position on the detector. The strength of all rays that impact the same pixel is totaled regardless of their phase. Using photometric units instead of radiometric ones results in Incoherent Illuminance.

Coherent Irradiance: This is the coherent power per unit area as a function of spatial position on the detector. To aggregate the complex amplitude of each ray striking the same pixel, the real and imaginary components are tracked individually, taking into account the beam's phase. To calculate coherent power, square the resulting amplitude. When photometric units are used instead of radiometric units, this option changes to Coherent Illumination.

Coherent Phase is the phase angle of the complex amplitude sum utilized in Coherent Irradiance.

Radiant Intensity: The power per solid angle in steradians as a function of the incidence angle on the detector.

When photometric units are used instead

of radiometric units, this option changes to Luminous Intensity.

Radiance (location Space) is the power per area per solid angle measured in steradians based on the detector's spatial location. This graphic is similar to the Incoherent Irradiance display, with data values divided by two pi; thus, it cannot simultaneously show variations in both angle and position space.

Radiance (Angle Space) refers to the power per area per solid angle measured in steradians based on the detector's incidence angle.

Incident Flux: The incident flux for detector volumes is the amount of light entering each voxel, measured in source units (watts, lumens, or joules).

Absorbed Flux: The incident flux for detector volumes is the amount of light entering each voxel, measured in source units (watts, lumens, or joules).

Flux vs. Wavelength: The spectrum distribution of flux incident on the detector for colored objects. The plot will appear as a bar chart. Using the Flux vs. Wavelength analysis, you may determine the spectral distribution of any item (except detector colors).

2.5 Applications Of Zemax Opticstudio

Lens Design: Design and analyze lenses for cameras, telescopes, microscopes, and other imaging systems. The software allows precise control over lens parameters, improving image quality and reducing optical aberrations.

Illumination Systems: Designing illumination systems for applications such as automotive headlights, projectors, and surgical lighting optimizes light distribution patterns and increases efficiency.

Spectroscopy: Capable of designing and analyzing spectrometers. Model complex spectrometer configurations, evaluate spectral resolution, and assess the impact of optical components on overall system performance.

Telescope Design: Astronomical and terrestrial telescope designing to create optical systems with exceptional imaging quality. The software helps minimize aberrations, optimize the field of view, and achieve high-resolution images.

Laser Systems: It plays an essential role in developing laser systems, including beam shaping, focusing, and collimation.

REFERENCES

1. https://www.zemax.com/
2. Katz, M. (2002). Introduction to geometrical optics. World Scientific.
3. Romano, A., & Cavaliere, R. (2010). Geometric optics. Springer.
4. Pecharsky, V. K., & Zavalij, P. Y. (2003). Fundamentals of diffraction (pp. 99-260). Springer US.
5. Naeem, M., Imran, T., Hussain, M., & Bhatti, A. S. (2022). Design simulation and data analysis of an optical spectrometer. Optics, 3(3), 304-312.
6. Naeem, M., Fatima, N. U. A., Hussain, M., Imran, T., & Bhatti, A. S. (2022). Design simulation of Czerny–Turner configuration-based Raman spectrometer using physical optics propagation algorithm. Optics, 3(1), 1-7.
7. Naeem, M., & Imran, T. (2022). Design and Simulation of Mach-Zehnder Interferometer by Using ZEMAX Optic Studio. Acta Scientific Applied Physics, 2(3).
8. Boreman, G. D. (2001). Modulation transfer function in optical and electro-optical systems (Vol. 4). Bellingham,

Washington: SPIE press.
9. Lin, P. D., & Liu, C. S. (2011). Geometrical MTF computation method based on the irradiance model. Applied Physics B, 102, 243-249.
10. Stamnes, J. J., & Heier, H. (1998). Scalar and electromagnetic diffraction point-spread functions. Applied optics, 37(17), 3612-3622.
11. Mahajan, V. N. (1983). Strehl ratio for primary aberrations in terms of their aberration variance, JOSA, 73(6), 860-861.

3. DESIGN SIMULATION OF OPTICAL SPECTROMETER

Spectrometers have a wide range of applications, from optical to non-optical spectroscopy. The need for compact, portable, and user-friendly spectrometers has been a focus of attention from small laboratories to the industrial scale. Spectrometers measure light of different wavelengths over a wide range of the electromagnetic spectrum. An Optical spectrometer is initially used as a scientific instrument that splits the beam into separate colors as an array, known as a spectrum. In the early days, Spectrometers were massive devices used in chemical laboratories to analyze large samples. Due to their advanced resolving power, diffraction gratings have replaced glass prisms in spectrometers. They are widely used to analyze material by spectroscopy. C. R. Masson developed a low-cost acoustic optical spectrometer for millimeter-wave observation with an effective resolution of 160 kHz[1]. Arzhantsev, S., and Maroncelli developed and characterized a spectrometer based on an optical Kerr shutter for emission gating and a polychromatic plus charge-coupled device (CCD) detection system for recording the time-resolved emission spectra of fluorescence species[2]. An acoustic-optical spectrometer was designed and tested for solar radio astronomy and variability studies of cosmic maser sources with a 5 m antenna (RT5)[3]. An imaging spectrometer based on the curved prism configuration has been modeled, showing the imaging quality factors influencing[4]. The imaging spectrometer is an optical instrument that simultaneously measures spectral and spatial properties. Because of its dimensional reliability and low cost, the prism-based dispersive spectrometer is one of the most used techniques in remote sensing. A micro-optical spectrometer in the 200–910nm spectral range was developed with a resolution of ~ 1nm[5]. An Offner spectrometer based on the geometrical study of ring fields was developed to improve the spectral resolution, and the analytical architecture was demonstrated using the optical design program Code V to construct a spectrometer with a range of 900 to 1700nm[6]. Gloria Mico presented the concept and design of an integrated optical device featuring evanescent field sensing and spectrometric analysis[7]. Such an integrated optics sensing spectrometer (IOSS) consists of a modified arrayed waveguide grating (AWG) whose arms are engineered into two sets with different focal points. Two reference designs have been provided for the visible and near-infrared wavelengths to determine the concentration of known solutes through absorption spectroscopy.

The Czerny–Turner Configuration-Based Raman Spectrometer using ZEMAX OpticStudio based on the physical optics propagation algorithm[8] is designed and simulated. In the simple optical spectrometer, there is no splitting of rays, and light interacts with only one optical component at a time in a particular sequence, so sequential mode can be used for designing and simulation.

3.1 General View Of An Optical Spectrometer

The Czerny–Turner configuration is one of the compact and flexible spectrometer designs, consisting of a single detector instead of a detector array and requiring fixed components, as shown in Figure 3.1.

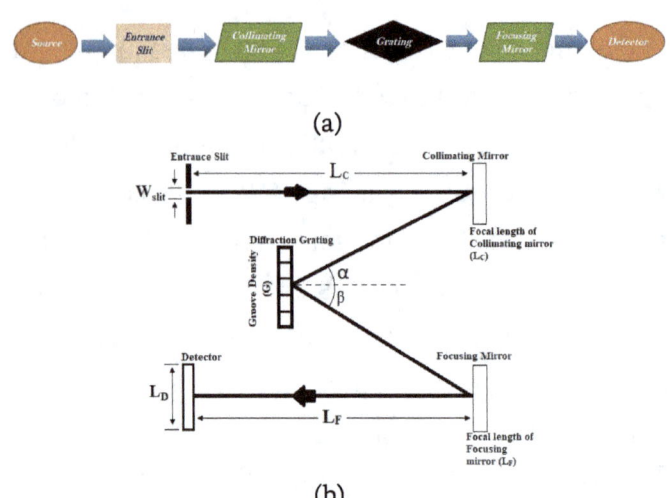

Figure 3.1 (a) Block diagram of the optical spectrometer. (b) Schematic representation of Czerny–Turner configuration-based spectrometer

The Czerny–Turner Spectrometer consists of one plane diffraction grating and two concave mirrors. The first mirror is a collimating mirror that collimates the beam and makes it equivalent to the grating surface of the spectrometer. The following mirror is the focusing mirror that focuses the input light from the source on the image sensor.

Initially, the wavelength range is selected, and the center wavelength is calculated as follows:

$$\lambda_2 - \lambda_1 = \Delta\lambda \quad ; \quad \lambda_c = \frac{(\lambda_2 - \lambda_1)}{2}, \quad (3.1)$$

If the deviation angle (θ) = 0, it will be the Littrow configuration[9]. According to the deviation, angles α and β are calculated.

$$\alpha = \sin^{-1}\left[\frac{\lambda_c G}{2}\cos\left(\frac{\theta}{2}\right)\right] - \frac{\theta}{2} \quad \beta = \theta - \alpha \quad (3.2)$$

Then, one must choose the detector size according to the size of the spectrometer. After selecting the detector size, the focal length of the focusing mirror L_F is obtained as,

$$L_F = L_D \frac{\cos(\beta)}{G(\lambda_2 - \lambda_1)} \quad (3.3)$$

The focal length of collimating mirror Lc is calculated as,

$$L_c = L_F\left[\frac{\cos(\alpha)}{\cos(\beta)}\right] \quad (3.4)$$

By using the collimating mirror L_c, the slit width is obtained as,

$$W_{slit} = G(\Delta\lambda)\frac{L_c}{\cos(\alpha)} \quad (3.5)$$

Here, G is the grating constant, and W_{slit} is the slit width, which is the entrance slit for any optical light. The careful selection of the slit width is an important factor when designing and developing a spectrometer[10].

3.2 Design And Simulation Of Optical Spectrometer

The design and simulations of the spectrometer were carried out using the ZEMAX OpticStudio[11] software. The parameters of each optical component defined in the lens data editor window were calculated per the mathematical model described in the previous section and listed in Table 3.1.

Surf: Type	Comment	Radius [cm]	Thickness [cm]	Semi-Diameter [cm]	Lines /μm
Standard	Source	Infinity	0.90	0.00	
Standard	Entrance slit	Infinity	4.55	0.08	
Standard	Collimating mirror	−10	−3.00	0.61	
Diffraction Grating	Diffraction grating	Infinity	3.00	0.52	0.600
Standard	Focusing mirror	−10	−4.55	0.61	
Standard	Detector	Infinity	-	0.30	

Table 3.1 Lens data editor window for the design of an optical spectrometer in ZEMAX OpticStudio

The table shows the surface type being set to standard except for the diffraction grating surface. The parameters in the lens data editor (LDE) were calculated according to the designed spectrometer. After defining the required surface and other relevant parameters involved in the design, the software provided a pictorial spectrometer layout, as shown in Figure 3.2, which included the source, input slit, collimating mirror, diffraction grating, focusing mirror, and image at the last surface which is a detector.

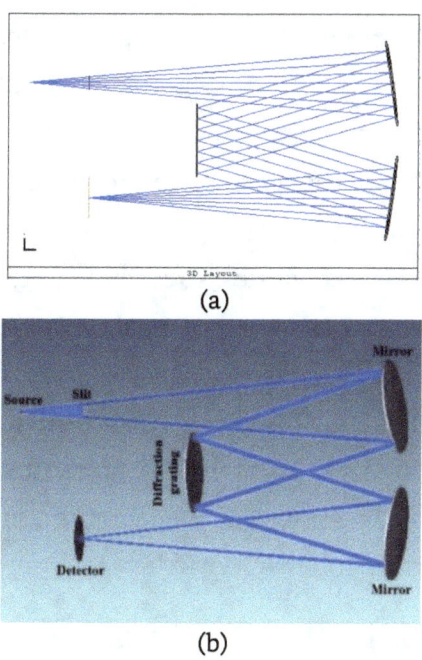

Figure 3.2 (a) 3-D layout. (b) Shaded model of the simulated design of the spectrometer in OpticStudio

The design and simulation of the spectrometer were carried out using the physical optics propagation (POP) algorithm for three different wavelengths: 0.5 μm, 0.6 μm, and 0.7 μm. The spectrometer simulation used a diffraction grating of 600 lines/mm. For the designed spectrometer image-space NA, object space NA, image-space F/#, total tracks, and stop radius, the values were 0.094, 0.088, 39.0625, 4.5, and 0.08 mm, respectively.

Zemax propagates any complex electric fields defined in three dimensions in POP mode. Typically, an ideal Gaussian beam defined by its waist, beam size, and relative z position propagates from interface to interface in the design. New set values such as waist, beam size, and position are saved at each interface. Hence, the properties of the pilot beam become very critical in determining whether the distribution of the propagating optical field is inside or outside the Rayleigh range and whether the algorithm is appropriate.

3.3 Simulation Results Of The Spectrometer In The Visible Region

Footprint Diagram: The footprint diagram shows when the light rays reach any optical object in front of their path and how they fall on that object's surface. This gives an idea of the number of rays reaching the object's surface and shows the pattern of falling on the surface.

When light fall on the slit, it blocks most of the light, and a small amount of the light passes through the slit. When light rays fall on the slit, all the rays of different wavelengths incident on the slit surface with the same pattern because they are coming from the same source with the same angle, so the footprint diagram of the entrance slit in Figure 3.3 (a) shows single wavelength pattern because other wavelengths are beneath that one. Figure 3.3 (b) shows the footprint diagram of the collimating mirror. When the light rays reach the collimating mirror, the wavelength pattern is the same as the entrance slit, but the intensity of the light decreases because the entrance slit blocks some of the light rays. The footprint diagram in Figure 3.3 (c) of the diffraction grating is different from the collimating mirror in orientation because collimating makes all the rays parallel, so some difference in the path of the rays occurs. Another reason for the change in orientation is that the collimating mirror is tilted at some angle, so it reflects the rays at some angle toward the grating, which causes changes in the path of the rays to occur.

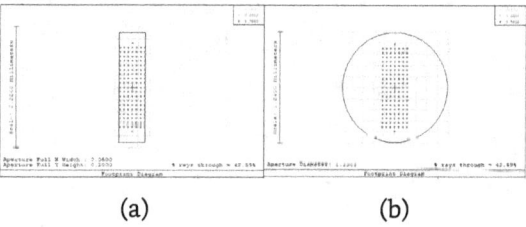

(a)　　　　　　　　　(b)

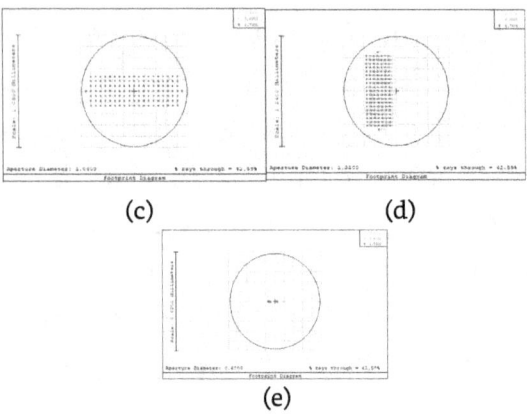

Figure 3.3 Footprint diagram of (a) Entrance slit (b) Collimating mirror (c) Diffraction grating (d) Focusing mirror (e) Detector

In Figure 3.3 (d), the three incident rays are now separated from each other and fall on the different points of the focusing mirror surface because diffraction grating reflects the different rays at different angles depending upon their wavelength. The orientation of the pattern of the focusing mirror is the same as the collimating mirror because the focusing mirror is tilted at some angle, so the rays coming from the grating change the orientation when they reach the mirror tilted at some angle. All three wavelengths are separated, so detecting differences in wavelength and intensity will be easy. Figure 3.3 (e) shows the footprint diagram of the detector where the image is formed. The detector is placed at the focal length of the focusing mirror, so the light rays coming from the focusing mirror will be focused on a single point. The figure shows the three different spots separated by minimal distance because the wavelengths are different, so they are separated from each other and can easily be analyzed now.

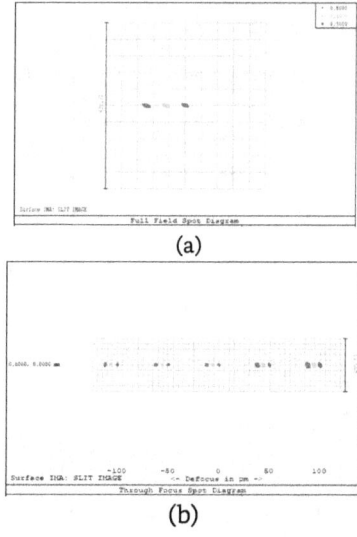

Figure 3.4 (a) Spot diagram at the detector of the designed spectrometer (b) Variation in the spot diagram relative to the distance of the detector from the focal point of the focusing mirror

Spot Diagram: The variation in the spot's size by changing the detector's position is shown in Figure 3.4. A focused spot will be obtained if the detector is placed at the mirror's focal point.

Figure 3.4 (a) shows the spot diagram of the wavelengths when the detector is at the focus point of the mirror. When the detector is towards or away from the focusing mirror, the spot size on the detector varies, as shown in Figure 3.4 (b). The target's geometrical representation becomes a finite blur spot in a decreased resolution when the detector is not in the paraxial region. As the separation between these planes increases, the blur spot becomes more prominent, and the resolution decreases further.

Optical Path Difference: The optical path difference is the relative path difference between the rays of different wavelengths passing through the medium. As the light moves through a medium, the frequency of the light remains the same, but its speed changes. So, when different wavelengths pass through the different mediums, some path differences appear that depend on the refractive index of the mediums and the distance traveled by the ray.

Optical Pathe Difference = $(n_1 * d_1) - (n_2 * d_2)$ (3.6)

n_1 and n_2 are the refractive indices of mediums, and d_1 and d_2 are the distance of ray passes through mediums 1 and 2, respectively[15].

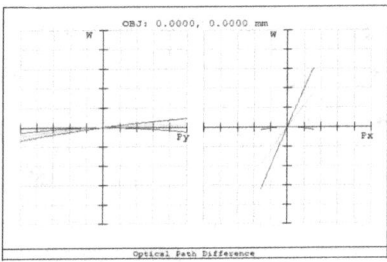

Figure 3.5 Optical path difference

Modulation Transfer Function (MTF): The Modulation Transfer Function (MTF) defines how much contrast is preserved by the detector in the original piece. Modulation Transfer Function (MTF) is an important method of describing the performance of an optical system; it determines how the spatial frequency is transmitted to the image. Suppose the pixels of the detector are significantly greater than the resolvable spot size. In that case, the optical system is said to be detector-limited, and the MTF of the system is reduced compared to the MTF the optical system itself can achieve.

MTF can be measured experimentally by imaging a small bar (or single-frequency sine) chart through the lens and onto the detector. The bar chart must be small because the lens's optical transfer function should not vary significantly over the target pattern. The range of MTF value is between zero to one. The spatial frequency where the MTF goes to zero is called the cut-off frequency. OTF stands for optical transfer feature. OLF of an optical device determines spatial frequencies managed by machines. The Modulation Transfer Function (MTF) is sometimes equivalent to OTF but ignores the phase effect.

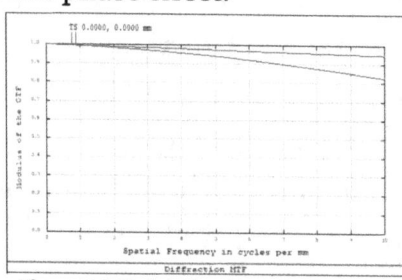

Figure 3.6 Modulation Transfer Function (MTF) of a designed spectrometer

MTF shows how the spatial frequency content of the entity is faithfully transmitted to the image and describes the performance of the optical system; the higher the value of MTF, the greater the device's image quality will be. As shown in Figure 3.6, when the spatial frequency is 10 lp·mm-1, the optical transfer function (MTF) is greater than 0.9, which means that the designed spectrometer system's image efficiency is higher.

Geometric Encircled Energy: Geometrical encircled energy calculates the concentration of energy in the optical picture. Figure 3.7 shows the variation in energy concentration when the image's size is changed on the detector. As the point image's size increases, the enclosed energy fraction increases.

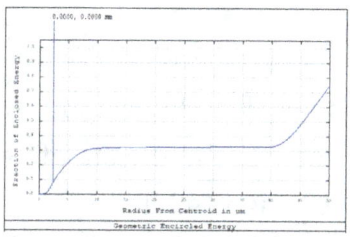

Figure 3.7 Geometric Encircled Energy

Relative Illuminance: Figure 3.8 shows the position of every wavelength falling on the detector.

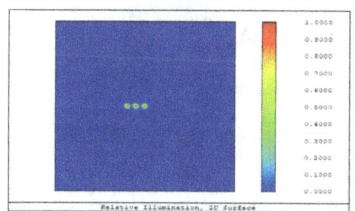

Figure 3.8 Relative Illuminance of the detector

All three wavelengths fall on different positions due to differences in wavelength; they are reflected at a different angle by the diffraction grating, so they are a little bit apart and focus at different points on the detector.

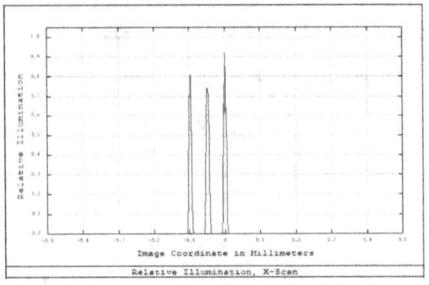

Figure 3.9 Relative illuminance along the x-axis of the detector

Figure 3.9 shows the relative illuminance in terms of X-coordinate. If the center of the detector is taken as the origin, then it is seen from the graph that one wavelength is focussed precisely at the origin, but the other two wavelengths move towards the negative x-axis and focus at the different points due to differences in the diffracted angle from the diffraction grating.

Spectral Irradiance: Irradiance is the power received by a surface per unit area. The SI unit of irradiance is the watt per square meter (W·m-2). The CGS unit erg per square centimeter per second (erg·cm-2·s-1) is often used in astronomy. Irradiance is also called intensity. Spectral irradiance is the irradiance of the surface per unit frequency or wavelength.

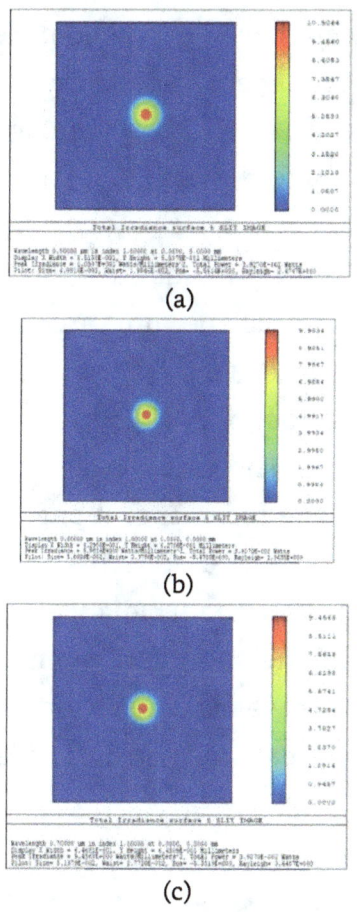

Figure 3.10. Images of the spectral irradiance on the detector at (a) $\lambda = 0.5\ \mu m$ (b) $\lambda = 0.6\ \mu m$ (c) $\lambda = 0.7\ \mu m$

The spot size at 0.5 µm, 0.6 µm, and 0.7 µm, respectively, in terms of total irradiance at the surface of the detector, is shown in Figure 3.8. Irradiance at the center was maximum because most of the rays were focused at the center;

moving away from the center decreased the irradiance, and energy per unit area delivered to the surface decreased[12,13]. The images of spectral irradiance showed minimal aberration in the optical range[14].

Figure 3.10 shows radiant flux (power) received by the detector surface per unit area and beam spot size at the detector surface for all wavelengths.

Irradiance across the X-Cross section: Figures 3.11 and 3.12 show the irradiance (intensity) variation in the focus spot along the horizontal (X-cross) and vertical (Y-cross) directions. The graph shows that the irradiance is maximum at the center of the spot, and when moving along a horizontal or vertical direction, the irradiance decreases symmetrically on both sides[16].

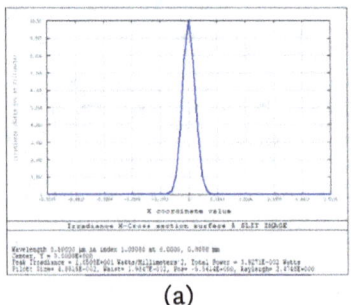

(a)

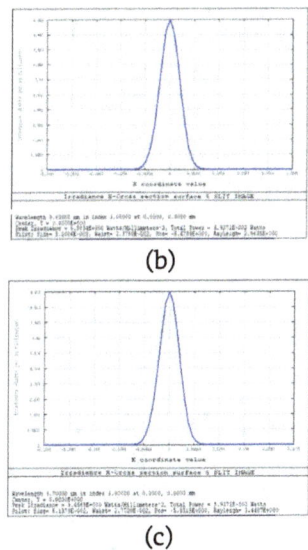

Figure 3.11 X-Cross irradiance of (a) λ=0.5 microns (b) λ=0.6 microns (c) λ=0.7 microns

Irradiance across the Y-Cross section: The spectrometer design is intended at visible wavelengths; therefore, the analysis is done primarily at 500nm and also observed at 600nm and 700nm using the POP algorithm. The results obtained demonstrate that there is no aberration and the overall design is accurate.

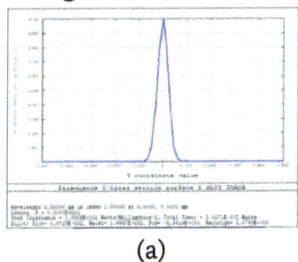

(a)

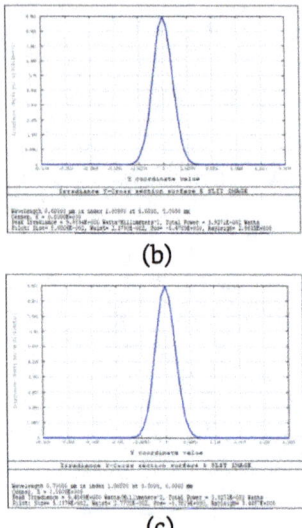

Figure 3.12 Y-Cross irradiance of (a) λ=0.5 microns
(a) λ=0.6 microns(a) λ=0.7 microns

REFERENCE

1. Masson, C. R. (1982). A stable acousto-optical spectrometer for millimeter radio astronomy. Astronomy and Astrophysics, Vol. 114, P. 270, 1982, 114, 270.
2. Arzhantsev, S., & Maroncelli, M. (2005). Design and characterization of a femtosecond fluorescence spectrometer based on optical Kerr gating. Applied spectroscopy, 59(2), 206-220.
3. Herrera-Martínez, G.; Luna, A.; Carrasco, L.; Shcherbakov, A.; Sánchez, D.; Mendoza, E.; Renero, F (2009). A Design of an Acous-to-Optical Spectrometer. Rev. Mex. De Astron. Y Astrofísica, 37, 156–159.
4. Feng, L., Wei, L., Yang, L., He, X., & Zhou, J. (2019, April). Modeling and Simulation of spectrometer based on prism. In Proceedings of the 8th International Conference on Software and Information Engineering (pp. 132-135).
5. Xu, D., Sui, C., Tong, J., & Yang, H. (2012, December). Optical Design of Micro Spectrometer. In 2012 Second International Conference on Instrumentation, Measurement, Computer, Communication and Control (pp. 339-341). IEEE.

6. Kim, S. H., Kong, H. J., Lee, J. U., Lee, J. H., & Lee, J. H. (2014). Design and construction of an Offner spectrometer based on geometrical analysis of ring fields. Review of Scientific Instruments, 85(8).
7. Micó, G., Gargallo, B., Pastor, D., & Muñoz, P. (2019). Integrated optic sensing spectrometer: Concept and design. Sensors, 19(5), 1018.
8. Naeem, M., Fatima, N. U. A., Hussain, M., Imran, T., & Bhatti, A. S. (2022). Design simulation of Czerny–Turner configuration-based Raman spectrometer using physical optics propagation algorithm. Optics, 3(1), 1-7.
9. Kao, C. F., Lu, S. H., Shen, H. M., & Fan, K. C. (2008). Diffractive laser encoder with a grating in Littrow configuration. Japanese Journal of Applied Physics, 47(3R), 1833.
10. Nagdive, A., Dongre, M., & Makkar, R. (2017, March). Design and simulation of NIR spectrometer using Zemax. In 2017 International Conference on Innovations in Information, Embedded and Communication Systems (ICIIECS) (pp. 1-5). IEEE.
11. Zemax (An Ansys Company); OpticsAcademy (Optics Studio). https://www.zemax.com/products/opticstudio
12. Osman, A. A., & Hassan, A. B. (2019, July). Simulation of hexagonal

segmented mirror of adaptive optics by using zemax program. In AIP Conference Proceedings (Vol. 2123, No. 1). AIP Publishing.
13. Bergström, E. T., Goodall, D. M., Pokrić, B., & Allinson, N. M. (1999). A charge coupled device array detector for single-wavelength and multiwavelength ultraviolet absorbance in capillary electrophoresis. Analytical chemistry, 71(19), 4376-4384.
14. Naeem, M., & Imran, T. (2022). Design and Simulation of Mach-Zehnder Interferometer by Using ZEMAX Optic Studio. Acta Scientific Applied Physics, 2(3).
15. Naeem, M., Imran, T., Hussain, M., & Bhatti, A. S. (2022). Design simulation and data analysis of an optical spectrometer. Optics, 3(3), 304-312.
16. Lan, G., & Li, G. (2017). Design of ak-space spectrometer for ultra-broad waveband spectral domain optical coherence tomography. Scientific reports, 7(1), 42353.

4. DESIGN SIMULATION OF RAMAN SPECTROMETER

Raman spectroscopy is a commonly used technique for analyzing the composition of substances. Its principle is to study the Raman scattering spectrum to obtain information on molecular vibration and rotation with a frequency different from the incident light, thereby judging the composition of substances. Light source laser has promoted the rapid development of Raman spectrometers since its appearance in the 1960s. The development of Raman spectrometers started relatively early in foreign countries. The leading companies in the United States include Nicolet. These companies have a high market share. There are relatively few domestic institutions specializing in developing and producing spectrometers. Zhejiang University and the Chinese Academy of Sciences have independently developed Raman spectrometers and applied them in scientific research. Currently, most Raman spectrometers adopt the Czerny-Turner structure based on the grating. Many novel designs of the Raman spectrometer have been published. Using convergent illumination of the grating for the fluorescence spectrum of organic particles, a photomultiplier tube, Wang et al. designed a crossed Czerny–Turner spectrometer. Ge et al. proposed a double-grating monochromator with a different fiber arrangement. We introduce the design simulation of the Raman spectrometer's optical structure using plain, reflecting grating and a focusing spherical mirror, shortening the overall system's length to obtain higher resolution, thus enabling the development of a robust and low-cost Raman spectrometer for optical measurements.

A typical spectrometer consists of a source light, an entrance slit, spherical collimating and focusing mirrors, grating, and CCD (charge-coupled device) detector[1]. Diffraction gratings are extraordinary due to their imaging efficiency in resolution and lowest wavefront aberration. However, they usually cannot achieve a flat focus curve[2], so a spherical mirror is needed to concentrate the subject on the image surface. Figure 4.1 (a) shows the schematic diagram representing the Raman spectrometer's optical configuration layout. Raman scattered light from the sample is focused on the entrance slit with the focusing lens of numerical aperture 0.16, further collimated with collimating mirrors and incident on the diffraction grating. The Raman signal is diffracted from the grating and splits into components focused on the CCD camera through the focusing mirror, as shown in Figure 4.1 (a). The radius of the curvature of the collimating and focusing mirror is R_1 and R_2, respectively.

A sample is usually defined before the entrance slit of the spectrometer. The sample's upper and lower energy levels have certain vibrational levels in an electronic state, which can be represented by horizontal lines in an energy state. The vibrational transition can occur when molecules move from one vibrational level to another energy level. One can identify and study the functional groups or chemical substances by monitoring the vibrational transition.

4.1 Mathematical Analysis Of Spectrometer

In Figure 4.1 (b), angles φ_1 and φ_2 are the diffraction angles of the two edge wavelengths after passing through the diffraction grating; they intersect the focusing mirror at A and B. σ_1, and

σ_2 are the angles between the reflected light of the two edge wavelengths on the focusing mirror and the horizontal line. Finally, they gather on points C and D of the linear CCD.

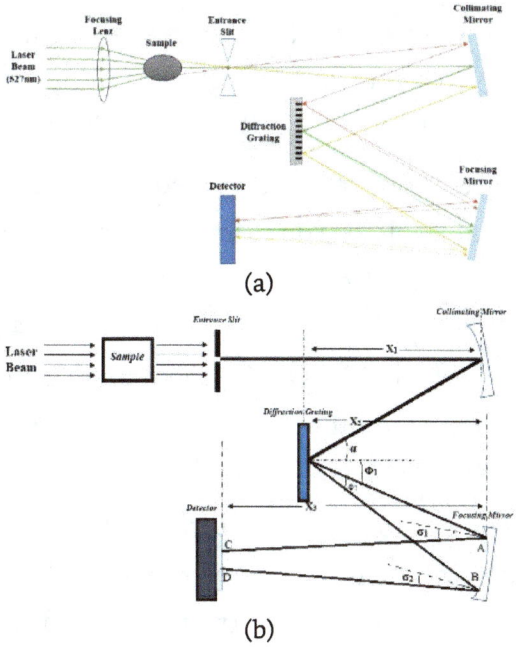

(a)

(b)

Figure 4.1 (a) Layout of Raman spectrometer: A laser beam is scattered from samples; this passes through the slit, is collimated by a mirror, and is directed to the diffraction grating, which resolves the wavelengths, which, in turn, are finally directed to the CCD by a focusing mirror.
(b) Analysis of the imaging plane in the Raman spectrometer

Using the diffraction grating formula,
$$d * \sin\theta = n\lambda \tag{4.1}$$
Where d is the grating constant, diffraction angles φ_1 and φ_2 can be calculated as:

$$d * (\sin(\alpha) + \sin(\phi_1)) = n\lambda_1 \quad (4.2)$$
$$d * (\sin(\alpha) + \sin(\phi_2)) = n\lambda_2 \quad (4.3)$$

Taking the diffraction order equal to 1 (n=1). Minimum size (length) of the focusing mirror to reflect the two edge wavelengths according to the geometrical structure:

$$AB = x_2 * [\tan(\alpha + \phi_1) - \tan(\alpha + \phi_2)] \quad (4.4)$$

After reflection from the focusing mirror, the horizontal angle of the diffracted rays is changed by 2θ as given below:

$$\sigma = \alpha + \phi - 2\theta \quad (4.5)$$

The size of the image is calculated using the geometrical structure:

$$CD = AB - x_3 * [\tan(\sigma_2) - \tan(\sigma_1)] \quad (4.6)$$

Equation 6 shows the minimum length of the CCD detector surface. The system's total function parameters can be determined based on the original configuration and the formula 4.1 to 4.6[3].

4.2 Design Parameters For The Simulation Of The Raman Spectrometer

The different designing parameters, such as each lens's surface type, curvature radius, thickness, and focal length, were chosen in the lens data editor window in Zemax[4,5], as shown in

Table 4.1. The physical optics propagation (POP) algorithm is adopted over geometrical ray tracing because geometrical ray tracing can only be used when diffraction limits are negligible.

Surface Type	Comment	Radius of Curvature (mm)	Thickness (mm)	Glass	Semi-diameter (mm)
Standard	Source	Infinity	0.90		0.00
Standard	Sample	Infinity	0.90		0.30
Standard	Slit	Infinity	4.50		0.08
Standard	Collimating mirror	-10.00	-3.00	MIRROR	0.61
Diffraction Grating	Grating	Infinity	3.00	MIRROR	0.52
Standard	Focusing mirror	-10.00	-4.50	MIRROR	0.61
Standard	Detector	Infinity	-		0.30

Table 4.1 Lens data editor

Values of different parameters like radius thickness are defined. Diffraction gratings are exceptional due to their imaging efficiency. However, they usually cannot achieve a flat focus curve[2]. The sample is defined before the entrance slit; upper and lower energy levels have certain vibrational levels. The sample absorbs the incident light and excites it to the upper energy level. Electrons in the excited state decay to a lower vibrational level through a non-radiative process and then decay to a lower energy level and emit photons of different energy, called Raman spectra.

After defining the required surface and other relevant parameters involved in the design, the software provided a pictorial spectrometer layout, as shown in Figure 4.2.

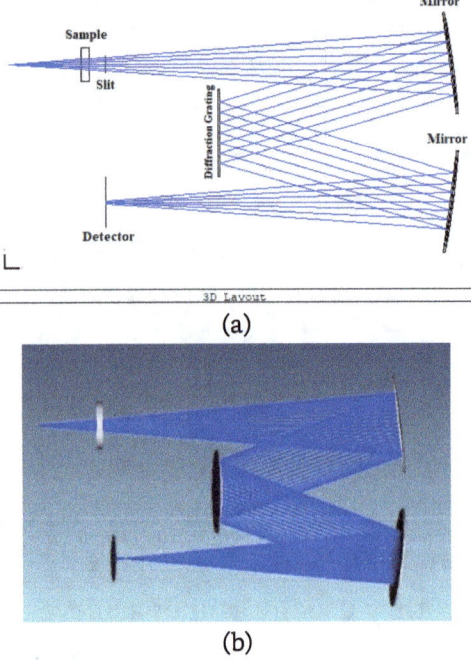

Figure 4.2 (a) 3-D Layout. (b) Shaded model of the simulated design of the spectrometer in OpticStudio

The initial parameters of the system design are shown in Table 4.2.

Parameter	Value
Laser Wavelength	527 nm
Observed Raman Spectra	(530 – 630) nm
Entrance Pupil Diameter	0.16 mm
Grating Constant	0.100 μm
CCD Pixel	1024 x 1024
Slit width	30 μm
Size of Pixel	(5 x 5) μm
R_1	-10 mm
R_2	-10 mm
x_1	3 mm
x_2	3 mm
x_3	4.2 mm

Table 4.2. Initial structure parameters of Raman spectrometer

4.3 Simulation Of Raman Spectrometer

Spot Diagram: After optimization, x_1 increases to 3 mm. The light diffracted by the grating has a certain convergence angle, corresponding to x_3, which is reduced to 4.5 mm, and other parameters like detector size and pixels and mirror focal length are unchanged. Figure 4.3 (a-c) shows the optimized spot diagram of the adjacent wavelengths. Figure 4.3 (d) shows the optimized spot diagram of the adjacent wavelengths at the edge and center wavelengths. An image formed on the spectrometer's plane shows that the adjacent wavelengths at different positions are well separated and distinguished[3,6].

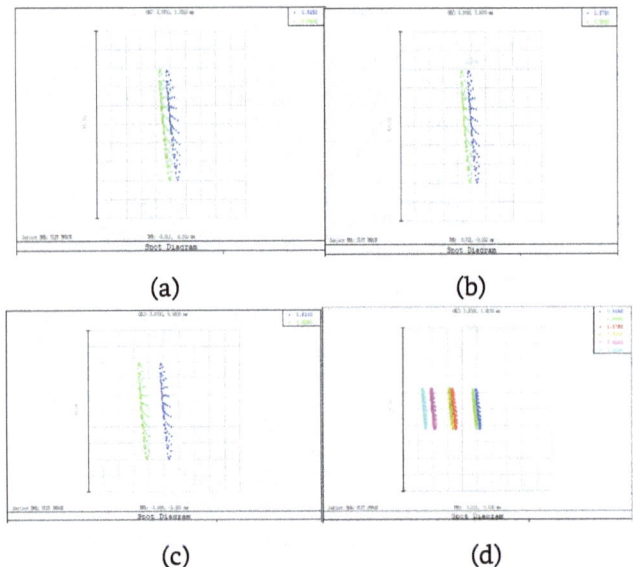

Figure 4.3 Spot diagram of (a) 535 (blue) & 540 nm (green) (b) 575 (blue) & 580 nm (green) (c) 610 (blue) & 625 nm (green) (d) Combine all wavelengths

Energy Distribution: Figure 4.4 (a-c) shows the energy distribution of adjacent wavelengths. Figure 4.4 (d) shows the energy distribution of all the different wavelengths expressed by the geometric linear response function. The linear response function is the cross-sectional representation of the image density distribution[7]. It can be seen in Figure 4.4 (d) that the energy spectrum of the neighboring wavelengths at various wavelengths can be easily distinguished.

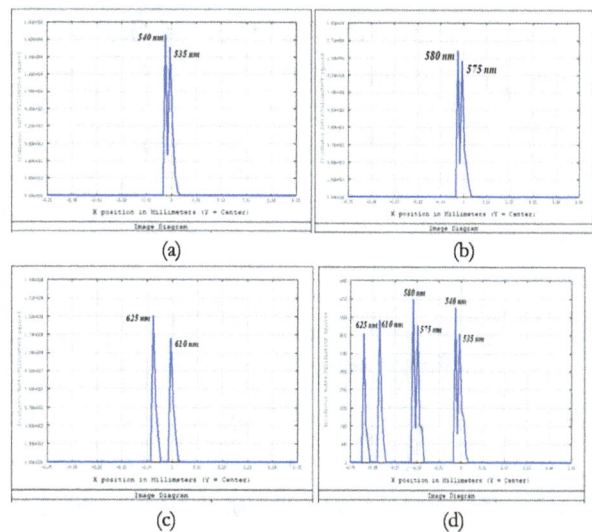

Figure 4.4 Energy distribution at (a) 535 & 540 nm (b) 575 & 580 nm (c) 610 & 625 nm (d) All wavelength

Modulation Transfer Function (MTF): The MTF diagram at the central wavelength (575 nm) shown in Figure 5 shows the contrast ratio between the input and output images. The MTF curve is approximately the same in the whole band, so the only curve of the center wavelength is given here. MTF shows how the spatial frequency content of the entity is correctly transferred to the image and describes the performance of the optical system. The higher the value of MTF, the greater the device's image quality. As shown in Figure 4.6, when the spatial frequency is 10 lp·mm-1, the optical (MTF) is greater than 0.8, which means that the efficiency of the designed spectrometer system is higher[8].

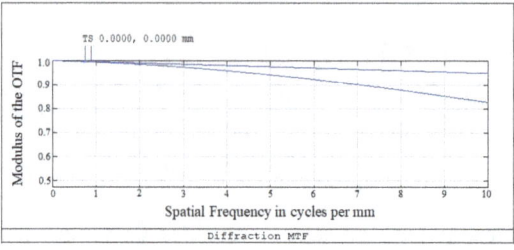

Figure 4.5 MTF curve at 575 nm

Geometric Encircled Energy: The optical term encircled energy calculates the energy concentration in the optical picture or predicted laser in a defined area. Energy concentration varies with the change in the size of the image on the detector. As the point image's size increases, the fraction of enclosed energy rises[9] as shown in Figure 4.6.

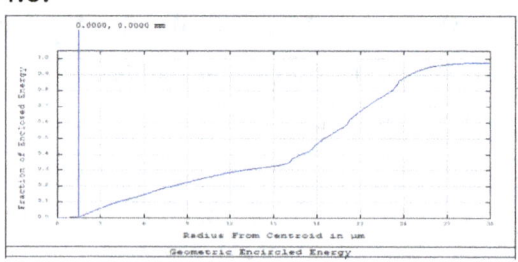

Figure 4.6 Geometric Encircled Energy

Irradiance and Relative Illuminance: The length and width of the receiving surface of the detector are 10 mm and 5 mm, respectively. The wavelength band imaged on the detector and the images produced by the two edge rays are similar to the two ends of the detector [Figure 4.7 (a)].

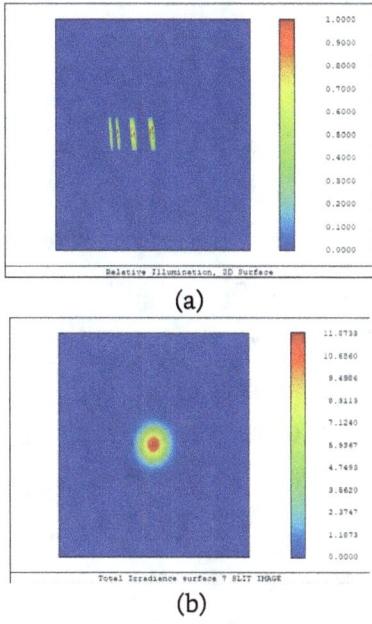

Figure 4.7 (a) wavelength band imaged on the detector plane.
(b) Irradiance and beam spot at the detector surface (580 nm)

Using the CCD, the light of the neighboring wavelengths of separate bands can also be separated, which indicates that the illumination of the adjacent wavelengths is easily separable and distinguishable from the present detector pixels. This finding further shows the functional viability of this system.

The radiant flux received by the detector surface per unit area and beam spot size at the detector surface for 580 nm wavelength is shown in Figure 4.7 (b).

$$I \; \alpha^1/_{\lambda^4} \qquad (4.7)$$

Illuminations at longer wavelengths result in a decrease in Raman signals. There is less illumination for a higher wavelength than the lower wavelength in the Raman spectrometer, so the above relation works correctly in the designed Raman spectrometer.

All results are obtained in the Sequential mode of Zemax. In this mode, the slit width can be measured only by increasing the height of the point light source. Follow the above settings to trace the light; the results obtained in this mode are closer to real-world observations.

REFERENCE

1. Tang, M., Fan, X., Wang, X., Xu, Y., Que, J., & He, J. (2015). General study of asymmetrical crossed Czerny–Turner spectrometer. Applied optics, 54(33), 9966-9975.
2. Lerner, J. M., Chambers, R. J., & Passereau, G. (1981, July). Flat field imaging spectroscopy using aberration corrected holographic gratings. In Imaging Spectroscopy I (Vol. 268, pp. 122-128). SPIE.
3. Naeem, M., Fatima, N. U. A., Hussain, M., Imran, T., & Bhatti, A. S. (2022). Design simulation of Czerny–Turner configuration-based Raman spectrometer using physical optics propagation algorithm. Optics, 3(1), 1-7.
4. Zemax (An Ansys Company); OpticsAcademy (Optics Studio). https://www.zemax.com/products/opticstudio
5. Naeem, M., & Imran, T. (2022). Design and Simulation of Mach-Zehnder Interferometer by Using ZEMAX Optic Studio. Acta Scientific Applied Physics, 2(3).

6. Naeem, M., Imran, T., Hussain, M., & Bhatti, A. S. (2022). Design simulation and data analysis of an optical spectrometer. Optics, 3(3), 304-312.
7. Li, Y., Oldendick, J., Ordonez, C. E., & Chang, W. (2009). The geometric response function for convergent slit-slat collimators. Physics in Medicine & Biology, 54(6), 1469.
8. Park, S. K., Schowengerdt, R., & Kaczynski, M. A. (1984). Modulation-transfer-function analysis for sampled image systems. Applied optics, 23(15), 2572-2582.
9. Everson, M., Duma, V. F., & Dobre, G. (2017, January). Geometric & radiometric vignetting associated with a 72-facet, off-axis, polygon mirror for swept source optical coherence tomography (SS-OCT). In AIP Conference Proceedings (Vol. 1796, No. 1). AIP Publishing.

5. DESIGN AND SIMULATION OF INTERFEROMETERS

Interferometers are investigative instruments used in a variety of scientific and technical disciplines. They were invented in the 1800s and are known as interferometers because they combine light sources to generate an interference pattern that can be measured and analyzed. Interferometers generate interference patterns that can reveal details about the sample or process being examined. They are frequently used to take extremely minute measurements that would otherwise be impossible to obtain. The interference concepts are simple to grasp. Waves interact with one another. The interference pattern is formed by combining the troughs and crests of the overlapping waves where they interact. Total constructive interference happens when the peaks and troughs of identical waves align perfectly, resulting in a larger wave whose height and depth equal the sum of the individual waves at each point of overlap. On the other hand, total destructive interference occurs when the peaks of one wave align perfectly with the troughs of another, causing them to cancel each other out and producing no wave.

When light from a single source is divided into two collimated beams and interferes can be seen with the detector, an interferometer measures the relative phase shift changes. Phase changes between two beams generated by a sample or a change in path length have been measured using the interferometer, among other things. In 1891, work by German physicist Ludwig Zehnder was modified by his son, Ludwig Mach, in a subsequent publication in 1892. The device is named after them[1-3]. In 1992, a sensor system was made to measure gaseous compounds using an integrated optical Mach-Zehnder interferometer (IO-MZ chip). By this means, the experimental signal obtained can be correlated to changes in the refractive index of the polymer layer. The dependence of the interference patterns on wavelength is explained[4]. It was first reported in 1998 that the Mach-Zehnder interferometer (MZI) might be used to analyze biomolecular surface multilayers. A three-waveguide coupler structure at interferometer output benefits signal reference, establishment, and maintenance of a sensitive operating point to the sensor design[5]. Based on non-reciprocal interference, a waveguide optical isolator has been developed. Ridge waveguides are constructed on a single sheet of bismuth-, lutetium-, and neodymium-iron garnet in a Mach–Zehnder configuration[6]. MZI-based logic gates were constructed using the efficient Mach–Zehnder interferometer (MZI) structure; the voltages on each electrode affect MZI's performance. The voltage fluctuation in the electrodes has been shown to alter electricity transmission[7].

There are different types of commercial interferometers available. The two basic types of interferometers are the Mach-Zehnder interferometer and the Michelson interferometer.

5.1 Mach-Zehnder Interferometer

It uses a monochromatic source, plane mirrors, beam splitters, and two detectors. The optical route lengths in the two arms may be substantially equal or vary. The block diagram of the Mach-Zehnder interferometer is shown in Figure 5.1.

The distribution of optical powers between the two outputs is determined by the precise difference in optical arm lengths as well as the wavelength. If the interferometer is correctly aligned, the path length difference may be changed so that the total power for a given optical frequency is directed to one of the outputs. For misaligned beams, fringe patterns will appear in both outputs, and adjustments in the path length difference will mostly impact the morphologies of these interference patterns, but the distribution of total powers on the outputs may not vary significantly.

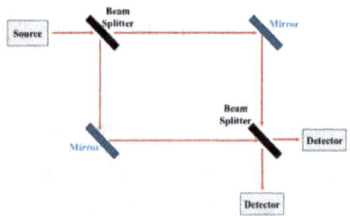

Figure 5.1 Block diagram of Mach-Zehnder interferometer

When employed as a tunable narrow-band filter, Michelson interferometers have several benefits and disadvantages over rival technologies like Fabry-Pérot interferometers. Michelson interferometers have the largest field of view for a given wavelength. They are relatively simple to operate, as tuning is accomplished through mechanical rotation of waveplates rather than high voltage control of piezoelectric crystals or lithium niobate optical modulators, as in a

Fabry-Pérot system. Michelson interferometers are less sensitive to temperature than Lyot filters, which employ birefringent components. On the downside, Michelson interferometers have a very limited wavelength range and need prefilters, which reduce transmittance[19].

The design simulation of the compact and small-size Mach–Zehnder interferometer is carried out using Zemax. Because of the small length of the overall system, it gives better resolution. Thus enabling the development of robust and low-cost Mach–Zehnder interferometer for optical measurements.

Mathematical Explanation: *Path Difference and Phase Shift:* The path difference (ΔL) between the two beams creates a phase difference ($\Delta \varphi = 2\pi \Delta L / \lambda$ where ΔL is the difference in path lengths of the two beams, λ is the wavelength of the light used.

Interference Condition: The intensity (I) of the interference pattern depends on the phase difference: $I = I_0(1+\cos(2\pi \Delta L/\lambda))$, where I_0 is the maximum intensity.

Constructive Interference ($\Delta\varphi = 2n\pi$): $I = 2I_0$

Destructive Interference ($\Delta\varphi = (2n+1)\pi$): $I = 0$

Design Parameters: The non-sequential component window in Zemax[8,9], shown in Table

5.1, defines the different designing parameters, such as the surface type, positions, angle, and Glass type. We opted for the physical optics propagation (POP) approach because geometrical ray tracing can only be employed when diffraction restrictions are insignificant[10,11].

Surface Type	X Position (mm)	Y Position (mm)	Z Position (mm)	Tilt About X	Tilt About Y	Tilt About Z	Glass
Source Rectangular	0	0	0	0	0	0	-
Polygon Object	0	0	20	-45.000	0	0	BK7
Rectangle	0	0	20	-45.005	0	0	MIRROR
Rectangle	0	0	120	-45.000	0	0	MIRROR
Polygon Object	0	0	120	-225.00	0	0	BK7
Detector	0	0	150	0.000	0	-90	ABSORB
Detector	0	0	120	90.000	0	-90	ABSORB

Table 5.1 Non-sequential component window in Zemax for Mach-Zehnder interferometer design

The system includes a He-Ne laser, two plane mirrors, two beam splitters, and two CCD (Charge-coupled device) detectors. The front faces of the beam splitter plates are coated (1:1) so that 50% of the light gets reflected and 50% is transmitted. A He-Ne laser is split at the first splitter and recombined by the second splitter. One of the mirrors is at a slight angle with respect to the second, so the beams are recombined with some tilt. Figure 5.2 (a) shows the schematic diagram and the layout for the Mach-Zehnder interferometer's optical configuration. Figure 5.2 (b) shows the shaded configuration model; the two detectors at the end show the interference pattern.

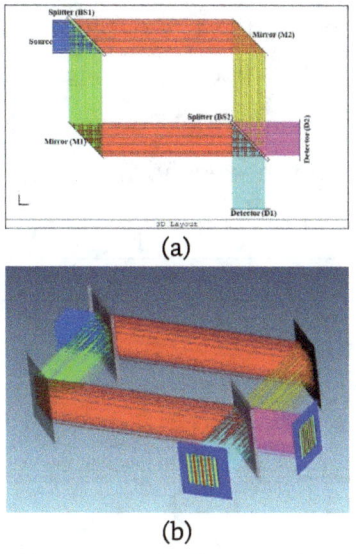

Figure 5.2 (a) 3D Layout of Mach-Zehnder Interferometer
(b) Shaded model of Mach-Zehnder Interferometer

We chose a 632.8 nm center wavelength laser as the light source, whose output power is 500 mW. A beam splitter (BS1) divides the incoming coherent light beam into two identical ones. Before they reach a detector, each of these beams takes a distinct path and is recombined. Their route lengths cause the phase difference between the two beams. The interference pattern between the originally identical waves is caused by the imposed phase difference[12]. As illustrated in Figure 5.1, both beams are redirected towards the second beam splitter (BS2), where they interfere, and the results are recorded at the detectors on both sides of the splitter. A 512 × 512 pixels CCD is chosen to record the spatial profile. The scale of each pixel is (10×10) μm, and the receiving surface of CCD is 2.1 cm^2.

Parameter	Value
Laser Wavelength	632.8 nm
Power	500 mW
Number of Layout rays	50
Number of Analysis rays	3000000
CCD Pixel	512 x 512
Size of Pixel	(10 x 10) μm

Table 5.2 Initial structure parameters of Mach-Zehnder Interferometer

It is possible to tell whether a beam has been divided into two separate beams by their phase difference. If the route length or the refractive index changes, this might be a physical change or a change in the refractive index. The initial parameters of the system design are shown in Table 5.2.

Simulation Results and Analysis: The simulated fringe pattern that occurs due to interference of the light beams at the detectors is shown in Figure

5.3.

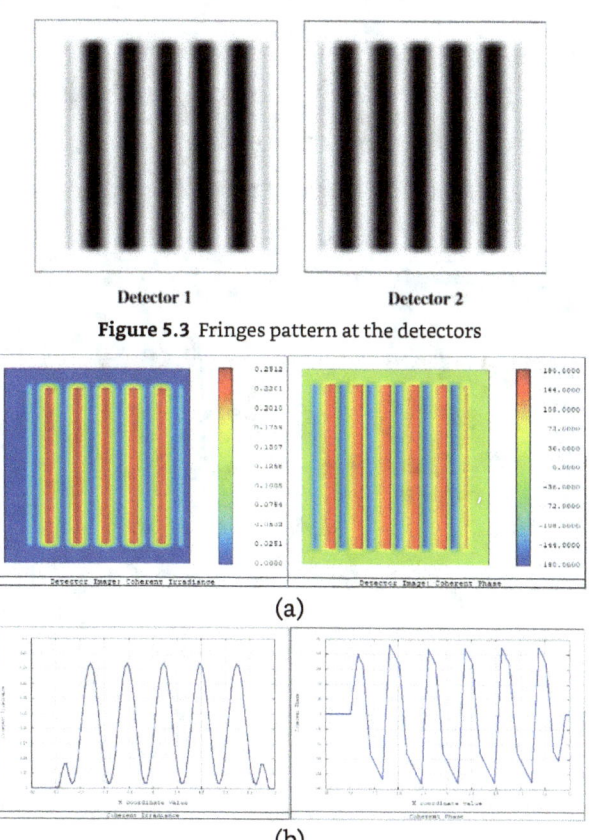

Figure 5.3 Fringes pattern at the detectors

(a)

(b)

Figure 5.4 (a) Spot diagram of the interference pattern (fringes) and coherent phase at detector 1 (b) Comparison of coherent phase and coherent irradiance at detector 1

The optimized spot diagram in the form of irradiance and phase is shown in Figure 5.4 (a). An image formed on the image plane shows that the beams interfere and fringes are well-separated and clearly defined. Figure 5.4 (b) shows that the beams are in phase; they interfere constructively. There is maximum irradiance bright fringe, and as the angle increases, the irradiance decreases. When the beams are out of phase, they interfere destructively, and the coherent irradiance goes to zero[10,13].

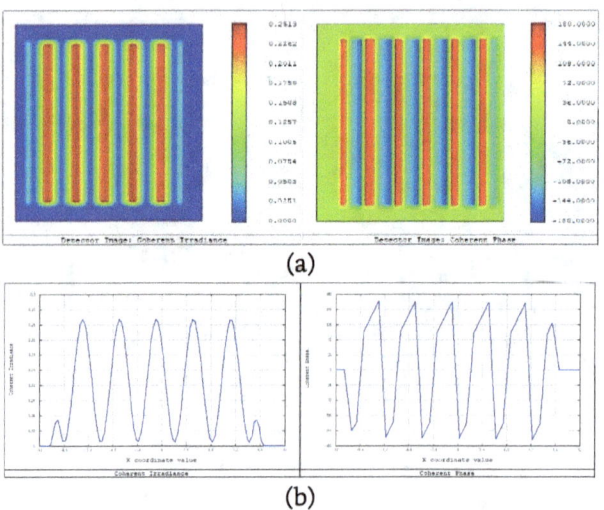

Figure 5.5 (a) Spot diagram of the interference pattern (fringes) and coherent phase at detector 2 (b) Comparison of coherent phase and coherent irradiance at detector 2

Figure 5.5 shows the results detected at detector 2. All results are obtained in the Non-Sequential Mode of Zemax. The coherent sum of the array amplitudes shows the interference fringes between the two beams. ZEMAX can model any form of the shearing interferometer using this technique. The total power is conserved between incoherent and coherent additions. Still, the peak irradiance of the coherent beam is two times that of the incoherent beam sum, as it should be[14,15]. It is possible to tell whether a beam has been divided into two separate beams by their phase difference. If the route length or the refractive index changes, this might be a physical change or a change in the refractive index. By introducing a sample in one of the paths, we can observe the interference pattern change at the detector, and from that, we can extract the information about that sample[16,17].

5.2 Michelson Interferometer

Albert Abraham Michelson developed the Michelson interferometer[18], which employs a single beam splitter to separate and recombine beams, as shown in Figure 5.6. If the two mirrors are oriented for precise perpendicular incidence, only one output is accessible, while the light from the other output returns to the light source. If the optical feedback is not desired and access to the second output is required, beam recombination might occur at a different position on the beam splitter. One option is to utilize retroreflectors, which make the interferometer relatively insensitive to minor misalignment of the retroreflectors. Simple mirrors with slightly different incidences can also be employed.

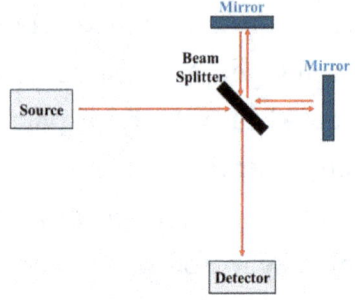

Figure 5.6 Block diagram of Michelson Interferometer

In the classic Michelson-Morley experiment, Michelson employed a broadband light source, necessitating the construction of an interferometer with almost no arm-length difference. The Michelson interferometer comes in a variety of configurations nowadays. A Twyman-Green interferometer, for example, is simply a Michelson interferometer with arms with extended beams. It measures the surface shape and transmitted wavefront quality of optical components, assemblies, systems, and optical-grade mechanical surfaces. A simple Michelson interferometer is simulated using ZEMAX with a rectangular light source.

Mathematical Explanation: The operation of the Michelson interferometer can be explained using basic wave interference principles. When two light beams of the same wavelength λ recombine, the path difference ΔL between the two beams determines the interference pattern.

Constructive Interference: Occurs when the path difference is an integer multiple of the wavelength, $\Delta L = m\lambda$, where m is an integer. This results in bright fringes.

Destructive Interference: Occurs when the path difference is an odd multiple of half the wavelength, $\Delta L = (m+1/2)\lambda$. This results in dark fringes.

The intensity I of the interference pattern at any point can be described by the equation: $I = I_0(1 + \cos(2\pi \Delta L / \lambda))$, where I_0 is the maximum intensity.

Surface Type	X Position (mm)	Y Position (mm)	Z Position (mm)	Tilt About X	Tilt About Y	Tilt About Z	Glass
Source Rectangular	0	0	0	0	0	0	-
Polygon Object	0	0	120	-45.006	0	0	BK7
Rectangle	0	30	120	-90	0	0	MIRROR
Rectangle	0	0	150	0	0	0	MIRROR
Detector	0	-50	120	90.000	0	0	ABSORB

Table 5.3 Non-sequential component window in Zemax for Michelson interferometer design

Design Parameters: The non-sequential component window in Zemax[8], shown in Table 5.3, defines the different designing parameters, such as the surface type, positions, angle, and Glass type for the Michelson interferometer. The system includes a rectangular laser, two plane mirrors, one beam splitter, and one CCD (Charge-coupled device) detector. The front faces of the beam splitter plates are coated (1:1) so that 50% of the light gets reflected and 50% is transmitted. Figure 5.7 (a) shows the schematic diagram and the layout for the Michelson interferometer's optical configuration. Figure 5.7 (b) shows the shaded model[8,9] of the configuration; the detector at the end shows the interference pattern.

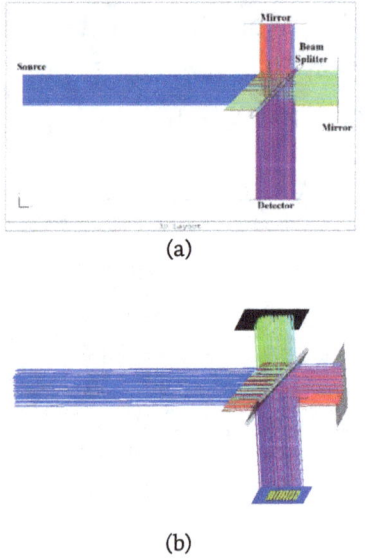

(a)

(b)

Figure 5.7 (a) 3D Layout of Michelson Interferometer
(b) Shaded model of Michelson Interferometer

Simulation Results and Analysis: The simulated fringe pattern that occurs due to interference of the light beams at the detector in the Michelson interferometer is shown in Figure 5.8.

Figure 5.8 Fringes pattern at the detector

The optimized spot diagram in the form of irradiance and phase is shown in Figure 5.9 (a). An image formed on the image plane shows that the beams interfere, and fringes are well-separated and clearly defined.

Figure 5.9 (b) shows that where the beams are in phase, they interfere constructively, then there is maximum irradiance (bright fringe), and as the angle increases, the irradiance decreases. When the beams are out of phase, they interfere destructively, and the coherent irradiance goes to zero[11].

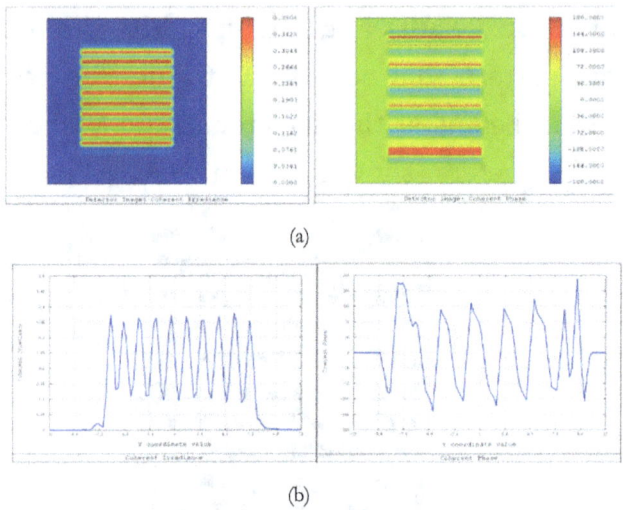

Figure 5.9 (a) Spot diagram of the interference pattern (fringes) and coherent phase at the detector. (b) Comparison of coherent phase and coherent irradiance at the detector.

The light with a center wavelength of 632.8 nm is used for simulation and optimization in the Zemax optical modeling program. Spot diagram, irradiance plot, and coherent phase are simulated. The analysis shows that the image efficiency of the system is higher, and interference patterns can be obtained using the above-designed Mach-Zehnder and Michelson Interferometer system. This study could pave the way for designing different interferometers for optical measurements.

5.3 Applications

Interferometers have a wide range of

applications beyond merely measuring length. Here are several examples:
- Gravitational wave detection using LIGO interferometer[20]
- Measuring rotations (with a Sagnac interferometer)[21]
- Measuring the linewidth of a laser[22]
- Reveals minor refractive index differences or induced index changes in a transparent material.
- For modulating the power or phase of a laser beam
- The full characterization of ultrashort pulses via spectral phase interferometry

REFERENCES

1. Bennett, F. D., & Kahl, G. D. (1953). A generalized vector theory of the Mach-Zehnder interferometer. JOSA, 43(2), 71-78.
2. Mach, L. (1892). On an interference refractor. Journal of Instrumentation, 12(3), 89.
3. Ji, Y., Chung, Y., Sprinzak, D., Heiblum, M., Mahalu, D., & Shtrikman, H. (2003). An electronic mach–zehnder interferometer. Nature, 422(6930), 415-418.
4. Fabricius, N., Gauglitz, G., & Ingenhoff, J. (1992). A gas sensor based on an integrated optical Mach-Zehnder interferometer. Sensors and Actuators B: Chemical, 7(1-3), 672-676.
5. Luff, B. J., Wilkinson, J. S., Piehler, J., Hollenbach, U., Ingenhoff, J., & Fabricius, N. (1998). Integrated optical mach-zehnder biosensor. Journal of lightwave technology, 16(4), 583.
6. Fujita, J., Levy, M., Osgood Jr, R. M., Wilkens, L., & Dötsch, H. (2000). Waveguide

optical isolator based on Mach–Zehnder interferometer. Applied Physics Letters, 76(16), 2158-2160.

7. Kumar, S., Raghuwanshi, S. K., & Kumar, A. (2013). Implementation of optical switches using Mach–Zehnder interferometer. Optical Engineering, 52(9), 097106-097106.

8. https://www.zemax.com/products/opticstudio.

9. Naeem, M., Fatima, N. U. A., Hussain, M., Imran, T., & Bhatti, A. S. (2022). Design simulation of Czerny–Turner configuration-based Raman spectrometer using physical optics propagation algorithm. Optics, 3(1), 1-7.

10. Naeem, M., & Imran, T. (2022). Design and Simulation of Mach-Zehnder Interferometer by Using ZEMAX Optic Studio. Acta Scientific Applied Physics, 2(3).

11. Naeem, M., Imran, T., Hussain, M., & Bhatti, A. S. (2022). Design simulation and data analysis of an optical spectrometer. Optics, 3(3), 304-312.

12. Zetie, K. P., Adams, S. F., & Tocknell, R. M. (2000). How does a Mach-Zehnder interferometer work?. Physics Education,

35(1), 46.

13. Rarity, J. G., Tapster, P. R., Jakeman, E., Larchuk, T., Campos, R. A., Teich, M. C., & Saleh, B. E. A. (1990). Two-photon interference in a Mach-Zehnder interferometer. Physical review letters, 65(11), 1348.

14. Li, L., Xia, L., Xie, Z., & Liu, D. (2012). All-fiber Mach-Zehnder interferometers for sensing applications. Optics express, 20(10), 11109-11120.

15. Chen, P., Shu, X., & Sugden, K. (2017). Ultra-compact all-in-fiber-core Mach–Zehnder interferometer. Optics letters, 42(20), 4059-4062.

16. Ahmed, F., Ahsani, V., Jo, S., Bradley, C., Toyserkani, E., & Jun, M. B. (2018). Measurement of in-fiber refractive index change using a Mach–Zehnder interferometer. IEEE Photonics Technology Letters, 31(1), 74-77.

17. Pan, S., & Yao, J. (2009). Switchable UWB pulse generation using a phase modulator and a reconfigurable asymmetric Mach-Zehnder interferometer. Optics letters, 34(2), 160-162.

18. Michelson, A. A. (1920). On the application of interference methods to astronomical measurements. Astrophysical Journal, vol. 51, p. 257, 51, 257.
19. Gary, G. A., Balasubramaniam, K. S., & Sigwarth, M. (2003, February). Multiple-etalon systems for the Advanced Technology Solar Telescope. In Innovative Telescopes and Instrumentation for Solar Astrophysics (Vol. 4853, pp. 252-272). SPIE.
20. Abramovici, A., Althouse, W. E., Drever, R. W., Gürsel, Y., Kawamura, S., Raab, F. J., ... & Zucker, M. E. (1992). LIGO: The laser interferometer gravitational-wave observatory. science, 256(5055), 325-333.
21. Culshaw, B. (2005). The optical fiber Sagnac interferometer: an overview of its principles and applications. Measurement Science and Technology, 17(1), R1.
22. Xue, J., Chen, W., Pan, Y., Shi, J., Fang, Y., Xie, H., ... & Su, B. (2016). Pulsed laser linewidth measurement using Fabry–Pérot scanning interferometer. Results in physics, 6, 698-703.

APPENDIX

This appendix is intended as a quick reference for the key features of ZEMAX OpticStudio.

Lens Data Editor

The Lens Data Editor is used to input and modify the parameters of optical components in your design. This includes adjusting curvatures, thicknesses, and materials of lenses, mirrors, and other elements.

Layout Plots

Layout plots visually represent your optical system, showing how light rays propagate. These plots are crucial for understanding your design's behavior and troubleshooting potential issues.

Analysis Tools

ZEMAX OpticStudio offers a wide range of analysis tools to evaluate the performance of your optical system. These include spot diagrams, wavefront maps, MTF (Modulation Transfer Function) analysis, and more.

Optimization Tools

Optimization tools help refine your design

by adjusting parameters to meet specific performance criteria. You can set goals and constraints, and ZEMAX will iteratively improve your design to achieve the desired.

Non-Sequential Mode

In Non-Sequential Mode, you can model complex optical systems where light can scatter, reflect, and refract multiple times. This mode is essential for simulating realistic scenarios like stray light analysis and illumination design.

User-Defined Surfaces

ZEMAX allows custom optical surfaces that can be defined by mathematical equations or imported from external data to be created. This feature is useful for designing unique optical elements that are

Tolerance Analysis

Tolerance analysis tools help you understand how manufacturing and alignment errors affect the performance of your optical system. This analysis is critical for ensuring your design is robust and reliable in real-world conditions.

Scripting and Extensions

ZEMAX supports scripting, allowing users to automate repetitive tasks and create custom functions. This capability extends the software's functionality and can significantly enhance productivity.

INDEX

1024X, 44
2D Layout, 28
Abbe V-numbers, 39
aberration, 4, 31, 32, 36, 37, 40, 50, 65, 68, 73
aberrations, 20, 28, 29, 34, 35, 40, 47, 48, 50
ABSORB, 45
Absorbed Flux, 47
absorption, 4, 5, 7, 13, 40, 53
achromatic, 2
acoustic-optical spectrometer, 52
activation, 26
adjacent wavelengths, 79, 82
Albert Abraham Michelson, 97
Albert Einstein, 2
Alhazen, 1
alterations, 14
amplitude, 46
amplitudes, 96
analysis, 16, 19, 20, 23, 28, 34, 35, 40, 41, 44, 47, 49, 52, 68, 71, 85, 101, 104
Analysis, vii, 16, 20, 28, 33, 34, 35, 42, 74, 75, 94, 100
analytical, 13, 14, 52
analyze optimal focus, 30
angle of incident, 1
angles, 7, 28, 37, 54, 59, 74, 75
antenna, 52

anti-nodal, 38
anti-principal, 38
aperture, 6, 73
applications, x, 7, 18, 47, 51, 102, 106
Arab world, 1
architecture, 52
Aristotle, 1
arm-length, 97
artifacts, 45
asteroids, 4
astigmatism,, 30
astronomy, 13, 16, 52, 65
automation, 5
bar chart, 37, 47, 62
bars, 23, 44
beam, 6, 11, 13, 14, 18, 36, 39, 40, 46, 48, 51, 54, 57, 66, 75, 83, 89, 92, 93, 96, 97, 99, 102
Beam File Viewer, 41
beam recombination, 97
beam splitter, 11, 13, 92, 93, 97, 99
BFS, 38
biomolecular, 88
birefringent, 90
BK7, 27
blaze angle, 7
blur, 61
blur spot, 61
bright fringes, 98
button, 23, 24, 44
CAD, 21
calculate, 12, 18, 46

calibration, 6
Cardinal Points, 38
CCD, 5, 8, 51, 73, 74, 75, 76, 82, 92, 93, 99
CGS unit, 65
characterization, 102
charge-coupled device, 51, 73
charged coupled device, 8
Christiaan Huygens, 1
CIE, 41
Coatings, 40
Code V, 52
coherent, 13, 40, 46, 93, 95, 96, 101
collimating, 4, 6, 7, 54, 56, 58, 59, 73
collimating mirror, 6, 54, 58, 59
Collimating Mirror, 6
collimating mirrors, 73
Color Chart, 41
color effects, 2
color map, 31
colors, 1, 47, 51
compact, 51, 53, 90
complementary metal-oxide-semiconductor (CMOS), 8
complex, xi, 18, 19, 44, 46, 48, 57
composition, 4, 72
computation, 38, 50
computer, 11
Concave, 6
concave gratings, 7
conjugate, 42
constructive interference, 12, 87

contour map, 33
contrast, 12, 62, 80
convergent, 72
convex, 4
coordinate, 28, 29, 38, 65
crests, 87
Cross Section, 32, 33
cross-sectional, 14, 79
cut-off frequency, 62
Czerny–Turner, 15, 49, 53, 73
Czerny–Turner Spectrometer, 54
dark fringes, 98
database, 45
deactivation, 26
decoding, 11
design, x, 4, 18, 19, 21, 43, 44, 52, 55, 56, 57, 68, 73, 77, 78, 88, 90, 91, 93, 99
Design Parameters, vii, 91, 98
destructive interference, 3, 12, 87
detector, 5, 6, 8, 11, 13, 23, 27, 44, 45, 46, 47, 53, 54, 56, 59, 60, 61, 62, 63, 64, 65, 66, 73, 76, 79, 81, 83, 88, 93, 95, 96, 99, 100, 101
Detector Viewer, 45
development, 72, 90
deviation, 54
diagnostic, 11
diagram, 5, 6, 10, 25, 30, 41, 53, 57, 58, 59, 60, 73, 79, 80, 89, 92, 94, 95, 96, 97, 99, 100, 101
dialog box, 25
diameter, 13, 14, 26
Diattenuation, 40

diffracted rays, 75
diffraction, 3, 5, 7, 8, 11, 20, 30, 31, 32, 35, 49, 50, 51, 54, 56, 57, 58, 59, 64, 65, 73, 74, 75, 76, 91
diffraction curve, 7
diffraction grating, 5, 7, 56, 73
Diffraction grating, 5
Diffraction Grating, 7
dimensionless slope, 38
dimensions, 13, 57
direct integration technique, 31
Directivity Plot, 41
dispersion, 2, 7, 39, 45
Dispersion, 7, 39
dispersive element, 4
dispersive material, 7
dispersive spectrometer, 52
displacements, 12
distort, 8
double-grating, 73
doublet, 29
Down keys, 29
dual nature, x, 2
dust, 14
Edge Spread, 32, 34
editor, 24, 25, 55, 56, 76
Editor Window, 27
efficiency, 35, 38, 47, 63, 73, 77, 81, 101
elastic, 9
electrical signal, 8
electricity transmission, 89
electromagnetic wave, 3

electromagnetism, 2
electronic state, 74
Electrons, 77
Element Drawing, 29
elements, 13, 19, 21, 27
ellipse, 39
emission, 1, 3, 4, 5, 13, 16, 51
Encircled Energy, 34
energy, 3, 9, 10, 34, 63, 65, 74, 77, 79, 81
Energy, 9, 79, 80, 81
energy state, 74
entrance slit, 55, 58, 73
Entrance Slit, 6
environmental, 19
Euler, 2
excitation, 4, 9
experimentally, 62
Export 2D DXF File, 42
Export IGES Line Work, 42
Extended Diffraction Image, 35
Extended Source, 34
Extra Data, 24
Extra Data editors, 24
extraordinary, 73
Fabry-Pérot, 90
Fabry-Pérot system, 90
FFT PSF, 32
fiber, 4, 6, 14, 35, 38, 73, 106
Fiber Coupling Efficiency, 38
filters, 11, 21, 90
fine-resolution, 4

fine-tune, 20
fluorescence, 52, 72
Flux vs. Wavelength, 47
focal length, 6, 36, 37, 54, 59, 76, 79
focal, principal, 38
focus, 6, 29, 30, 31, 34, 51, 61, 64, 65, 67, 73, 77
focusing mirror, 5, 8, 54, 56, 59, 60, 61, 73, 74, 75
Footprint Diagram, 36, 57
formula, 75, 76
Foucault's knife-edge shadowgrams, 33
Fourier transform, 13
fraction, 9, 63, 81
Franciscan Roger Bacon, 1
frequency, 4, 5, 7, 9, 10, 12, 30, 32, 61, 62, 63, 65, 72, 81, 90
Fresnel, 43
fringe, 90, 94, 95, 100, 101
Full Field, 30
functional groups, 74
functionality, 28, 42
galaxies, 4
gas, 14
Geometric Bitmap Image, 35
Geometric Encircled Energy, 63, 64, 81, 82
Geometric Line, 34
geometrical representation, 61
geometry, 7, 12, 33
glass, 24, 27, 51
Glass Map, 39
Gloria Mico, 52
gradient, 39

graph, 5, 31, 39, 40, 65, 67
graphical user interface, 18
grating, 4, 7, 8, 11, 27, 52, 54, 55, 56, 57, 58, 59, 64, 65, 72, 73, 74, 75, 78
grating groove frequency, 7
gravitational waves, 14
greyscale, 31
Grin Profile, 39
groove density, 7
groove facet angle, 7
grooves, 7
heating, 4
He-Ne laser, 92
highest-precision, 14
high-precision, 20
high-voltage, 5
horizontal, 67, 74, 75
human eye, 1
identical waves, 13, 87, 93
illuminance systems, 21
illumination, 21, 35, 36, 41, 42, 44, 47, 72, 82, 83
Illumination 2D Surface, 42
Illuminations, 83
image, 19, 20, 24, 29, 34, 35, 38, 47, 54, 56, 57, 59, 62, 63, 73, 76, 79, 81, 94, 100, 101
Image Simulation, 34
image-space, 57
imaging, 16, 18, 21, 23, 44, 47, 48, 52, 62, 73, 75, 77
imperfections, 12
impulse, 8

Incident Flux, 47
incident photon, 9
incoherent, 13, 46, 96
industry, 12, 18
infrared, 2, 52
Insert Lens, 23
instruments, 87
integrated, 39, 52, 88
intensity, 5, 12, 16, 40, 41, 58, 59, 65, 66, 91, 98
interaction, 9, 10
interface, xi, 23, 26, 43, 44, 57
interference, 3, 11, 14, 20, 87, 88, 90, 91, 92, 93, 94, 95, 96, 98, 99, 100, 101, 105
interference fringes, 13, 14, 96
interference pattern, 87
Interferogram, 33
Interferometry, vii, 1, 11
Internal Transmittance, 39
interpolation technology, 44
intersect, 74
invented, 4, 87
invention, 3
inverse, 7
investigative, 87
ion, 4
ions, 5
Irradiance, 46, 65, 66, 67, 81, 83
irradiance plot, 101
Kepler, 1
laser, 2, 11, 48, 72, 75, 81, 92, 93, 99, 102, 105, 106
Laser interferometry, 14

lasers, 4
Last Configuration, 26
layout, 29, 56, 73, 77, 92, 99
LDE, 26, 27, 56
LDT, 41
lens, 1, 18, 22, 23, 24, 25, 28, 29, 36, 42, 43, 47, 55, 56, 62, 73, 76
Lens Data, 23, 24, 26, 27
lenses, 1, 18, 19, 21, 24, 42, 43, 47
library, 19, 21
Library of Optical Components, 21
light, v, 1, 3, 4, 5, 6, 7, 8, 9, 10, 11, 12, 16, 18, 20, 21, 39, 43, 47, 51, 53, 54, 55, 57, 60, 61, 72, 73, 74, 77, 78, 82, 83, 87, 91, 92, 93, 94, 97, 98, 99, 100, 101
light source, 11, 12, 83, 93, 97
Light Source, 35
LightningTrace Control, 44
LIGO interferometer, 102
linewidth, 102, 106
Littrow configuration, 54
load data, 25
Ludwig Mach, 88
Ludwig Zehnder, 88
machines, 63
Mach–Zehnder configuration, 88
Mach-Zehnder interferometer, 12, 16, 88, 89, 91, 92
Mach–Zehnder interferometer, 89
Mach–Zehnder interferometer, 90
maintenance, 88

manufacturing, 19, 29
maser, 52
materials, 4, 9, 21
mathematical model, 55
matrix, 30, 41
Matrix, 30
maximum, 30, 65, 67, 91, 95, 98, 101
measurable phenomena, 8
measure, 12, 14, 51, 88
mechanical drawing, 29
mediums, 61
memory, 29
Menus, 23
Merit Function, 24
mesh, 44
Michael Faraday, 2
Michelson interferometer, 12, 89, 97, 98, 99, 100
micro-optical spectrometer, 52
microscopes, 19, 47
minimal, 60, 65
minimal distance, 60
minute, 12, 87
mirror, 4, 6, 7, 8, 11, 13, 27, 54, 56, 58, 59, 60, 61, 73, 74, 75, 79
mirrors, 6, 18, 19, 21, 54, 73, 89, 92, 97, 99
misalignment, 97
modern applied optics, 1
modes, xi, 9, 18, 21, 22
modulating, 102
modulation transfer function, 30, 31, 32
Modulation Transfer Function, 62, 63, 80

molecules, 9, 74
monochromatic source, 11, 89
monochromator, 4, 73
MTF, 25, 28, 30, 31, 32, 44, 49, 62, 63, 80, 81
Multi-Configuration, 24
multiple surfaces, 19
NA, 57
nanoscale, 4, 13, 14
Narcissus contribution, 38
narrow-band, 90
nature of light, x, 1, 2, 3
N-BK7, 27
Newton, 1, 3
Next Configuration, 26
Nicolet, 72
nodal, 38
noise, 8
non-optical, 51
non-sequential, 18, 21, 22, 43, 44
Non-Sequential mode, 23, 44
obscuration decenters, 28
observations, 83
Offner, 4, 52
one surface, 24
OpenGL graphics, 29
Opens an existing Zemax archive, 22
optical arm lengths, 89
optical astigmatism, 40
optical bench, 6
optical configuration, 73, 92, 99
optical design software, 18

optical measurements, 73, 91, 102
optical modulators, 90
optical path, 6, 13, 29, 61
Optical Path, 29, 61
optical path lengths, 13
optical power, 6, 37
optical powers, 89
optical range, 65
optical shop, 29
Optical spectroscopy, 3
optical system, x, 4, 6, 20, 26, 35, 40, 62, 63, 81
optical system parameters, 20
optical systems, xi, 4, 18, 19, 20, 21, 26, 48, 49
optical technology, 1
optical transfer function, 35, 62, 63
optics, 1, 3, 18, 19, 20, 49, 50, 52, 53, 56, 76, 91
OpticStudio, xi, 26, 55, 56, 78
optimization, 19, 20, 23, 78, 101
Optimization algorithms, 20
optimization tool, 19
optimized spot diagram, 79
orientation, 34, 39, 41, 58, 59
origin, 1, 65
outputs, 89
overlapping waves, 11, 87
parallel, 6, 58
parameters, 19, 20, 26, 40, 47, 55, 56, 76, 77, 78, 79, 91, 93, 94, 98
Paraxial Gaussian Beam, 40
paraxial region, 61
Partially Coherent Image, 35

particle, 2, 5
particles, 3, 5, 72
path difference, 13, 29, 61, 62, 91, 98
path length difference, 89
pattern, 12, 13, 41, 57, 58, 59, 62, 87, 91, 92, 93, 94, 95, 96, 98, 99, 100, 101
patterns, 3, 13, 44, 47, 87, 88, 90, 101
peak, 8, 96
performance, 6, 18, 19, 20, 21, 45, 48, 62, 63, 81, 89
phase, 12, 13, 33, 40, 46, 63, 88, 91, 93, 94, 95, 96, 100, 101, 102
Phase Aberration, 40
phase difference, 12, 13, 91, 93, 96
phenomenon, 8, 9
photoelectric effect, 3
photometers, 4
photometric, 46
photomultiplier, 72
photon, 9, 10
photonics, 18
photons, 3, 5, 9, 77
physical, 13, 20, 43, 49, 53, 56, 76, 91, 93, 96
physical optics, 20
Physical Optics Propagation, 16, 40
physics, 4, 106
picture analysis, 23, 34
piezoelectric crystals, 90
pilot beam, 57
pixel, 46, 93
plane, 7, 27, 30, 33, 40, 54, 75, 79, 83, 89, 92, 94,

99, 100
plasma, 5
Plato, 1
Plots, 33, 34, 39
plotting, 30
Polar Plot, 41
Polarization, 39
polarization-induced, 40
polychromatic, 51
POP, 56, 57, 68, 76, 91
portable, 51
power, 6, 7, 37, 39, 46, 47, 51, 65, 66, 89, 93, 96, 102
predicting, 20
Preferences, 23
prefilters, 90
principal wavelength, 26
prisms, 21, 43, 44, 51
programming, 18, 19
propagated rays, 21
properties, 9, 25, 52, 57
pupil, 25, 29, 36, 37, 38, 39, 40
Pupil Aberration, 29
pupil locations, 25
radiant flux, 66, 83
Radiant Intensity, 46
Radiant Source Model, 41
radiation transducer, 8
radio interferometry, 14
radiometric, 46
radius, 23, 26, 33, 34, 38, 42, 57, 73, 76, 77

radius of curvature, 23, 27
Raman, vii, xi, 8, 9, 10, 11, 15, 16, 49, 53, 72, 73, 75, 76, 77, 78, 83
Raman effect, 9, 16
Raman spectrometers, xi, 72
Raman spectroscopy, 9
ray, 18, 19, 20, 21, 25, 29, 31, 34, 35, 36, 38, 39, 40, 43, 44, 45, 46, 61, 76, 91
Ray Aberration, 29
Ray Database Viewer, 45
Ray Sampling, 44
Ray Trace, 38, 39
ray tracing, 18, 20, 35, 39, 43, 44, 76, 91
Rayleigh, 8, 9, 11, 57
recompute, 25
rectangular, 31, 32, 34, 98, 99
reflecting grating, 73
reflection, xiii, 1, 7, 40, 75
Reflection gratings, 7
refraction, 1, 39
refractive index, 12, 61, 88, 93, 96, 102
Relative Illuminance, 64, 81
relative phase difference, 12
research, 12, 14, 72
resolution, 4, 6, 7, 13, 34, 44, 48, 51, 61, 73, 90
retroreflectors, 97
RGB, 35
RGB bitmap file, 35
RMS, 33, 34, 42
Robert Kirchoff, 4
Robert Wilhelm Bunsen, 4

rotational, 9, 28
rotationally symmetric, 38, 40
RSMX, 41
Sag Table, 38
sagittal, 31, 33, 40
Sagnac interferometer, 102, 106
samples, 4, 51, 75
Saves the *.ZMX, 22
scale, 30, 51, 93
Scatter Function Viewer, 41
Scattered Light, 8
scattering, 8, 9, 10, 16, 21, 72
scientist, 1
scientists, 3
screen, 14
Semi-diameter, 27
sensing spectrometer, 52
sensitive, 88, 90
sensitivity, 11
sensor, 54, 88
sequential, 18, 21, 22, 23, 43, 53, 91, 98, 99
sequential mode, 18, 21, 23, 44, 53
Sequential mode, 21, 22, 23, 44, 83
Shaded Model, 29
shearing interferometer, 96
shift, 3, 9, 13, 30, 36, 39, 88
SI unit, 65
simulation, x, 16, 19, 49, 53, 56, 71, 72, 73, 76, 85, 90, 101, 104
singlet, 29
Skew Gaussian, 40

Skew Gaussian Beam, 40
slit, 4, 6, 54, 55, 56, 57, 59, 73, 74, 75, 77, 83
software, xi, 5, 18, 19, 20, 21, 26, 44, 47, 48, 55, 56, 77
solar system, 4
Solid Model, 42
Source Illumination, 41
source,, 27, 56, 89, 93, 97
space, 2, 47, 57, 71
spatial, 32, 45, 46, 52, 62, 63, 81, 93
species, 11, 52
spectral, 5, 6, 7, 8, 11, 41, 47, 48, 52, 65, 66, 71, 102
Spectral irradiance, 65
Spectral Irradiance, 65
spectral resolution, 6, 7, 52
spectrometer, 4, 5, 6, 7, 8, 10, 11, 16, 48, 49, 51, 53, 54, 55, 56, 60, 63, 68, 71, 72, 73, 74, 75, 76, 77, 78, 79, 81, 83, 85, 104
spectrophotometer, 11
spectroscope, 4
spectroscopy, 3, 7, 8, 9, 10, 11, 13, 51, 72
Spectroscopy, 4, 47
spectrum, 5, 47, 51, 72, 79
Spectrum Plot, 41
speed, 45, 61
spider obscurations, 28
spot, 19, 20, 25, 28, 29, 30, 33, 34, 42, 44, 60, 62, 65, 66, 67, 79, 83, 94, 100
spot diagrams, 19, 20, 28, 29, 30, 44
Spot Diagrams, 29
spreadsheet, 23, 24, 26

stability, 11
stars, 4
Stokes, 10
straight-line propagation, 1
Strehl ratio, 32, 33, 34, 50
structure, 14, 16, 72, 75, 76, 78, 88, 94
subject, 73
sunlight, 9
superposition, 11
surface, 5, 12, 14, 21, 22, 23, 24, 25, 27, 29, 31, 33, 34, 36, 38, 39, 40, 42, 43, 54, 57, 58, 59, 65, 66, 73, 76, 77, 81, 83, 88, 91, 93, 98
Surface Curvature, 33
Surface Phase, 33
Surface Sag, 33
symmetry, 28
system, 1, 4, 6, 9, 18, 20, 21, 23, 24, 26, 29, 36, 45, 48, 51, 62, 63, 73, 76, 78, 81, 82, 88, 90, 92, 93, 99, 101
System Menu, 24
tabulate, 38
tangential, 31, 33, 40
telescope, 1, 13, 48
temperature, 90
textual data, 28
theory, xi, 1, 3
thick element, 27
thickness, 24, 26, 39, 76, 77
Thomas Young, 2, 3
Through Focus, 30, 31
throughput, 6, 19, 20

Toggles, 22
tolerance analysis, 19
Tolerance analysis, 20
Tolerance Data, 24
tolerancing, 23
transmission, 7, 40
Transmission Fan, 40
transmission grating, 7
transmittance, 39, 90
transmitted, 10, 40, 62, 63, 92, 98, 99
transparent, 9, 102
troughs, 87
tunable, 90
two surfaces, 24
Twyman-Green interferometer, 98
types, 5, 7, 89
ultrashort pulses, 102
ultraviolet, 2
Universe, 14
variation, 60, 63, 66
vertical direction, 67
vibration, 8, 72
vibrational levels, 74, 77
vibrational transition, 74
virtual workspace, 18
visible, 4, 52, 68
visible range, 4
voltage, 89, 90
waist, beam size, 57
Wang, 72
wave, 1, 2, 3, 12, 51, 87, 98, 102, 105

wave theory, 1, 3
wavefront, 19, 20, 32, 33, 34, 37, 73, 98
wavefront aberrations, 19
wavelengths, 4, 7, 11, 25, 29, 51, 57, 58, 59, 61, 64, 65, 66, 68, 74, 75, 79, 80, 82, 83
wavelets, 32
waveplates, 90
weights, 26
white light, 11
Wille Broad Snell, 1
Wireframe, 42
X-cross, 67
Y-coordinate, 38
Y-cross, 67
YNI Contributions, 38
Zemax, 18, 21, 22, 23, 25, 26, 28, 29, 41, 57, 70, 76, 83, 85, 90, 91, 96, 98, 99, 101
Zemax LLC, 18
Zemax OpticStudio, 26
Zemax.CFG, 23
ZRD, 45

ACKNOWLEDGEMENT

I begin by offering my heartfelt thanks to Almighty ALLAH, the most benevolent and merciful, for creating a universe filled with perfect symmetry and harmony. I am deeply indebted to my parents for their steadfast support and encouragement. I want to express my deepest gratitude to everyone who supported and guided me throughout the creation of this book. I would also like to extend my heartfelt thanks to my mentor and co-author, Dr. Tayyab Imran, whose guidance, wisdom, and unwavering support were instrumental in creating this book. Your dedication, insights, and expertise have enriched every chapter, and your passion for the subject matter has been a constant source of inspiration. This book would not have been possible without your mentorship, and I am honored to have worked alongside you. Thank you for believing in me and being an exceptional mentor and co-author. Your expertise and guidance have been invaluable, and I deeply appreciate the time you invested in my learning. Special thanks to my fellow students and colleagues for their contribution, efforts, and collaboration. Your contributions, feedback, and shared passion for optical engineering made this project both enriching and enjoyable. Lastly, I am profoundly grateful to my family for their unwavering support and belief in me. Your encouragement gave me the strength and motivation to complete this project. This book reflects your love and support as it is of my efforts. Thank you for being my foundation.

ABOUT THE AUTHOR

Muddasir Naeem

He is pursuing his Ph.D. in Physics at Lahore University of Management Sciences (LUMS), Lahore, Pakistan. He has completed his BS (Hons) in Physics with a gold medal from the University of Wah, Pakistan. He has done his Master's in Physics from COMSATS University Islamabad, Pakistan. The author specializes in Optics, Laser, Condensed Matter Physics, and Magnonics. His areas of interest include Optical instrumentation, Laser development, Ultrashort laser, non-linear optics, Raman spectroscopy, and Brillouin light scattering spectroscopy.

Email: muddasirnaeem98@gmail.com

ABOUT THE AUTHOR

Tayyab Imran

Laser physicist specializing in ultrashort pulse laser technology, working at Extreme Light Infrastructure-Nuclear Physics facility, Romania. Ph.D. from the Korea Advanced Institute of Science and Technology, South Korea. Post-doctoral research at Instituto Superior Técnico, Portugal. He was an assistant professor at King Saud University, Saudi Arabia. As an associate professor, he has also served at COMSATS University Islamabad and Lahore University of Management Sciences, Pakistan. He taught, researched, and established laser laboratories at various universities. His research interests are related to laser engineering and technology. He is a recipient of several research fellowships.
Email:tayyabmrn@gmail.com
http://tayyabimran.weebly.com

BOOKS BY THIS AUTHOR

Laser Technology Made Easy: Building A Sealed Tube Carbon Dioxide (Co2) Laser System From Scratch For Students And Hobbyists

This short book focuses on developing a sealed tube carbon dioxide (CO2) laser from scratch with a reasonable depth without attempting to be overly broad, a mix of theory and experimental aspects. We start by looking at the fundamental concepts of laser, like population inversion, pumping scheme, feedback system, and active medium, and further move to the working of the CO2 laser, how the CO2 laser is electrically pumped and stabilized using the homemade cooling system.

It is helpful for readers who want to work on developing lasers with limited resources and is useful for graduate and undergraduate students taking a course on lasers and light sources. Hobbyists who want to work on developing lasers and have a basic understanding of the laser may get the most out of this book.

Guide To Building A Laser: Air-Based Nitrogen Laser Construction

Guide to Building a LASER is a meticulously

crafted resource, offers a clear and concise exploration of TEA nitrogen lasers. This book combines theory and practicality to provide a well-rounded understanding of the subject. Join us on this illuminating journey, where each chapter is thoughtfully designed to contribute to your deeper comprehension of the marvels of laser science and technology. Unveiling the marvels of air as a lasing medium and employing the Blumlein cavity configuration, this guide caters to seasoned scientists, aspiring students, and curious minds alike. Beginning with the fundamental principles of gas and TEA nitrogen lasers, the book intricately delves into the Blumlein cavity configuration, its design, and operational nuances. It demystifies high-voltage pulse generation and provides a detailed roadmap for inexpensively constructing a TEA nitrogen laser system. Offering insights into testing, tuning, and practical applications, the guide ensures optimal performance, accompanied by thorough safety guidelines. Aspiring researchers, engineers, and enthusiasts will find this book a guiding light into the fascinating world of TEA nitrogen lasers, merging theory, practical insights, and real-world applications. Embark on a journey into the future of laser science and technology with this compelling and professional resource.